Managing the Transition from Print to Electronic Journals and Resources

A Guide for Library and Information Professionals

Routledge Studies in Library and Information Science

Managing the Transition from Print to Electronic Journals and Resources

A Guide for Library and Information Professionals

Maria D. D. Collins, BA, MAT, MSLS
Patrick L. Carr, MA, MLS
Editors

Routledge
Taylor & Francis Group

NEW YORK AND LONDON

First published 2008
by Routledge
270 Madison Ave, New York, NY 10016

Simultaneously published in the UK
by Routledge
2 Park Square, Milton Park, Abingdon, Oxon OX14 4RN

Routledge is an imprint of the Taylor & Francis Group, an informa business

Transferred to Digital Printing 2009

Library of Congress Cataloging in Publication Data
   Managing the transition from print to electronic journals and resources: A guide for library and information professionals/Maria D.D. Collins, Patrick L. Carr, editors.
      p. cm.
   1. Libraries—Special collections—Electronic information resources. 2. Libraries—Special collections—Electronic journals. 3. Electronic information resources—Management. 4. Serials control systems. 5. Libraries and electronic publishing. 6. Academic libraries—Collection development—Case studies. 7. Digital libraries—Collection development—Case studies. 8. Academic libraries—Effect of technological innovations on. I. Collins, Maria D. D. II. Carr, Patrick L.
   Z692.C65M367 2008
   025.17'4-dc22

                                                                2007039740

ISBN 10: 0-7890-3336-4 (hbk)
ISBN10: 0-2038-8941-X (ebk)

ISBN 13: 978-0-7890-3336-9 (hbk)
ISBN 13: 978-0-2038-8941-1 (ebk)

CONTENTS

ABOUT THE EDITORS

Maria D. D. Collins, MSLS, MAT, BA, is Assistant Head of Acquisitions at the library at North Carolina State University.

Patrick L. Carr, MSLS, MA, BA, is Assistant Professor and Coordinator of Serials at the library at Mississippi State University.

CONTRIBUTORS

Beth Ashmore is a cataloging librarian at Samford University Library in Birmingham, Alabama. Here, she gets to work with serials, non-music media, and just about anything else that seems to defy classification. Before turning to the world of technical services, she worked in reference/instruction librarians positions at Samford and Mississippi State University. She is happy to know what things look like from both sides of the library aisle. She enjoys reading, writing, and filmmaking, where she works almost exclusively in the genre of library-related films.

Elizabeth S. Burnette joined the staff at the North Carolina State University Libraries in 2000. She is presently the interim head of the Acquisitions Department, responsible for overseeing daily operations, administering the Libraries' materials budget, and negotiating licenses. She obtained her MIS from Drexel University in 1990 and has worked as a law librarian for the Third Circuit Court of Appeals Library in Philadelphia, Pennsylvania and as a systems analyst and trainer for both the Third Circuit and the Administrative Office for the U.S. Courts in San Antonio, Texas.

Joan Conger, MLIS, MA.OD, coordinated electronic resources in a diverse range of academic libraries, combining reference, acquisitions, assessment, and training to create successful management of technology change during the 1990s and into this century. To share what she learned from these experiences, she wrote the textbook *Collaborative Electronic Resource Management* published by Libraries Unlimited. Joan is now pursuing a PhD in human and organization development from Fielding University.

Hilary Davis has been with North Carolina State University Libraries since 2005 as a fellow and is currently collection manager for physical sciences, engineering, and data analysis. She has also been engaged with the development of a digital repository for the University as part of the Libraries' strategic initiatives. She holds an MLS from University of Missouri-Columbia, an MS in biology (focusing on plant systematics) from University of Missouri-St. Louis, and a BS in Biology from Guilford College. She is an active member of the Special Libraries Association and keeps close ties with the botany research community (seedplants.org at the Missouri Botanical Garden).

Mark Ellingsen is the manager of the Web, Database and Library Systems team at the University of Bristol, United Kingdom. The team supports library applications including a library management system, meta-search software, and digital repository, as well as a bespoke electronic resource management system. The team also has responsibility for the University course management system, corporate Web infrastructure, and database services. Mark joined the University of Bristol in 1998. He has an MSc in computer science, a postgraduate diploma in librarianship and information science, and fifteen years' experience of managing IT support teams within the higher education sector. He was also involved in the development of Verde, the electronic resource management system from Ex Libris.

Christine L. Ferguson has been the electronic resources/serials librarian at Furman University since August 2003. Prior to assuming her position at Furman, she was the serials librarian at Mississippi State University. She holds a MS in information sciences from the University of Tennessee and a BA in Theatre Arts from the University of Richmond. Cris is a member of NASIG, ALA, and ACRL, and currently serves as an associate editor for *Against the Grain.* Cris and her husband John Larkin, a high school science teacher, share their home with three dogs, two cats, and one turtle.

Nancy Fried Foster is lead anthropologist and comanager of the Digital Initiatives Unit for the River Campus Libraries, University of Rochester. She is responsible for work-practice studies of faculty members, students, and library employees and for ensuring that members of these user groups participate in the libraries' codesign process. Nancy has extensive experience in anthropological research in the Amazon and Papua New Guinea, as well as in higher education in the United States and the United Kingdom.

Jill E. Grogg serves as the electronic resources librarian at The University of Alabama (UA) Libraries, and at UA, she is responsible for e-resource licensing and acquisition as well as federated search and link resolver administration. Jill has written widely about linking and the OpenURL, including authoring a *Library Technology Report* for ALA TechSource. She is a regular contributor to *Searcher: The Magazine for Database Professionals.* Jill was selected as a 2007 Mover & Shaker by *Library Journal.*

Selden Durgom Lamoureux is the electronic resources librarian at the University of North Carolina at Chapel Hill. Selden has been the principle license negotiator for UNC-Chapel Hill for the past seven years. She has conducted several licensing workshops as well as presented at a number of conferences on issues concerning electronic resource acquisition. Selden is an adjunct professor at the UNC-Chapel Hill School of Library and Information Science. She is currently involved in the Alternatives to Licensing initiative and is a member of the NISO SERU Working Group.

David Lindahl is director of digital library initiatives for the River Campus Libraries, University of Rochester. He is responsible for the advanced development of library systems including a faceted search and browse catalog interface and the institutional repository for the University of Rochester. A computer scientist, David has extensive experience with user-centered design and library-related digital research and development projects.

Bonnie Parks is head of the serials cataloging section at Massachusetts Institute of Technology Libraries and is the "Serials Conversations" column editor for the journal *Serials Review.* Bonnie enjoys teaching and has conducted numerous training sessions for the Serials Cataloging Cooperative Training Program (SCCTP). She has also taught the ALCTS Rules and Tools for Cataloging Internet Resources workshop.

Charley Pennell is principal cataloger for metadata at North Carolina State University (NCSU) Libraries, where he spends much of his time contemplating database structures and various descriptive metadata schemas. He has enjoyed a rich and varied career in cataloging, in both the United States and Canada, and has served as head of cataloging at Memorial University of Newfoundland, Villanova University, and NCSU. He holds an AB in Geology from Earlham College, where he worked as a student assistant under Evan Ira Farber and Tom Kirk, and an MLS from the University of Toronto, where he studied cataloging with Nancy Williamson and Margaret Cockshutt.

Jaroslaw Szurek is a music librarian and cataloger at Samford University Library in Birmingham, Alabama. A native of Poland, he worked at the Institute of Musicology Library, Jagiellonian University in Krakow, Poland, before coming to the United States. He has a master's degree in musicology with specialization in early music from the Jagiellonian University and a master's degree in information science from the University at Albany, SUNY.

Bonnie Tijerina is the electronic resources coordinator in the Collection Development Department at Georgia Institute of Technology's Library & Information Center. Bonnie earned her MLS from the University of Wisconsin, Madison and began her career as a North Carolina State University Libraries fellow. She is the coordinator of the Electronic Resources & Libraries (ER&L) Conference, a forum for information professionals to explore ideas, trends, and technologies related to electronic resources and digital services.

Jennifer Watson began her career at Oxford University in England. After gaining her postgraduate diploma at the University of North London, she headed to Los Angeles, California, where she held a variety of positions with library vendors, as well as volunteering with several nonprofit organizations. Since 2003, she has worked at the University of Tennessee Health Science Center in Memphis, Tennessee, first as electronic services librarian, and more recently as head, Electronic & Collection Services.

Jeff Weddle is an assistant professor in the School of Library and Information Studies at the University of Alabama. His primary areas of teaching are reference services for the twenty-first century, public libraries, collection development, and the American literary small press. His book, *Bohemian New Orleans: The Story of the Outsider and Loujon Press,* was published by the University Press of Mississippi in June 2007.

Glen Wiley is the digital resources and metadata librarian at Rensselaer Polytechnic Institute. He leads the library-wide selection, acquisition, licensing, maintenance, and access of electronic information resources and management tools. Glen also coordinates cataloging and metadata activity in both print and electronic formats. Prior to his appointment at Rensselaer Polytechnic Institute, Glen was a serials and electronic resources cataloging librarian at North Carolina State University Libraries and a visual resources cataloger at Syracuse University Libraries. He has presented on cataloging, metadata, and technology topics at various national conferences and has served as an online instructor for SOLINET (Southeastern Library Network). He received his BFA in history of art and an MSLIS from Syracuse University.

Foreword

It is a privilege to write the foreword for this book. Maria D. D. Collins and Patrick L. Carr and all the chapter authors have worked hard to put this monograph together. What strikes me particularly is how timely the information is, and how well the chapters represent the various aspects of managing the transition we are going through. This book provides an excellent snapshot of the period we are now experiencing.

The book is a "soup to nuts" smorgasbord that captures every aspect of the transition. I was pleased to see chapters on standards as well as topics one would expect, such as licensing and ERM implementation. A chapter that discusses institutional repositories was not a surprise to me, but might be to some, since there seems to be a struggle within the profession to determine who exactly will be managing IRs. It's obvious that the transition from print to electronic resources requires us to think differently about all the traditional roles in technical services operations.

The chapters devoted to case studies are welcome and useful because they represent diverse types of academic environments—everything from a community college to large research institutions is showcased. This inclusive array demonstrates that some of the most creative thinking is occurring at many kinds of institutions and is not necessarily limited to ARL libraries.

A number of our up and coming serials experts are represented in these pages. They are doing groundbreaking work and are paving the way for completing a transition that is exciting, frustrating at times, but never dull. It's wonderful to have such company in a time of such change!

Eleanor I. Cook
Serials Coordinator and Professor
Appalachian State University

Managing the Transition from Print to Electronic Journals and Resources

Acknowledgments

This book has been a true collaborative effort bringing together opinions and expertise from a wide range of authors. Patrick and I could never have achieved the scope of this work without the help of these authors, who we would like to thank for all of their time, effort, and wonderful contributions. Likewise, the completion of this work would not have been possible without the guidance and enthusiasm of Jim Cole, co-editor of The Haworth Press's series on serials and continuing resources. Along with these individuals, we would like to acknowledge the continued support of both Mississippi State University Libraries and North Carolina State University Libraries.

For their assistance during the final stages of the book's preparation, we are indebted to a number of individuals. We wish to extend a special thanks to Liz Burnette, who offered her enthusiasm, support, and organizational magic to the manuscript review and formatting process. In addition, we would like to thank Jacquie Samples and Robert Brown for their efforts to format chapters and Wayne Jones for his invaluable editorial assistance.

Finally, I would like to express my sincere gratitude to my husband, Leonard, and two sons, Aidan and Christopher, for sacrificing many an evening and weekend to this project.

Introduction

Libraries today are in a period of transition. As patrons' usage of print resources dwindles and as their demand for seamless and unbounded access to electronic resources continues to grow, extraordinary challenges are emerging. Indeed, success in today's information environment requires libraries to take dramatic steps in order to forge the partnerships and implement the tools and workflows that are appropriate for managing and providing access to materials that are increasingly acquired in electronic formats. In many instances, these steps force libraries to undergo a philosophical shift in how their collections are defined. Despite the enormity of these challenges, however, librarians equipped with the right knowledge and expertise have unprecedented opportunities to reimagine and redefine their libraries' strategies for effectively managing and providing access to resources; as T. Scott Plutchak has stated, "the great age of librarianship is just beginning."[1]

The goal of this book is to enable its readers to enter this new age of librarianship. In other words, the book aims to provide librarians who currently have or will have e-resource management and access responsibilities with the knowledge they need to understand the dynamics of their changing environment and thereby design, implement, and manage solutions that will enable their libraries to make a successful transition from collecting print resources to providing online access to e-resources. To achieve this goal, the book consists of eighteen chapters divided into four Parts, titled: "Evolving Collections," "Evolving Staff and Partnerships," "Evolving Tools," and "Evolving Strategies and Workflows." As their titles suggest, each of these sections focuses on fundamental areas of academic libraries that are evolving to shape libraries' transition to e-resources. In addition, each section provides a wealth of information, examples, and perspectives presented by authors currently working to address these challenges.

The transitions that libraries are currently experiencing are ultimately rooted in the changing formats of their collections. Part I, "Evolving Collections," explores these roots through chapters that examine how libraries are making the transition to electronic formats while addressing the implications that these formats have on efforts to effectively acquire and preserve content. Chapter 1, authored by Elizabeth S. Burnette, explores how the increasing centrality of e-resources is leading libraries to reconfigure existing acquisition and budgeting practices while experimenting with new acquisition models. Christine L. Ferguson builds on Burnette's discussion in Chapter 2, which discusses the myriad of factors that a library must take into account in order to develop a collection of e-resources that meets patrons' needs while maximizing fiscal resources. Of course, a library's successful selection and acquisition of e-resources are of limited value if the library cannot maintain access to these resources over a sustained period of time. In Chapter 3, author Jennifer Watson explores the topic of preservation in an information environment dominated by e-resources, highlighting the evolving network of players and initiatives that have arisen to overcome libraries' e-resource preservation challenges. Concluding the first section of the book is Hilary Davis' (Chapter 4) case study of how specific libraries are examining their unique characteristics and resources in order to successfully address the challenges that have come about as a result of their evolving collections. In particular, Davis spotlights the acquisition models of four libraries, two that have implemented wholesale shifts to online only collections and two that are exploring models in which patrons are provided with access to e-resources through pay-per-view purchases.

Part II, "Evolving Staff and Partnerships," focuses on the relationships that libraries have developed to effectively manage and provide access to their evolving collections. In Chapter 5, co-authors Joan Conger and Bonnie Tijerina outline the evolving models of communication, adaptation, and collaboration that libraries must foster in order to enable personnel to develop the skills and outlooks that fully reflect the dynamic nature of e-resources. Drawing from the results of a survey of ASERL libraries, Chapter 6, written by Maria D. D. Collins, builds on Conger and Tijerina's analysis by discussing the trends that are currently shaping the backgrounds and skill sets of library personnel being assigned e-resource responsibilities. In Chapter 7 of the "Evolving Staff and Partnerships" section, co-authors Beth Ashmore and Jaroslaw Szurek move beyond partnerships that exist among library personnel in order to examine how the emergence of e-resources has led librarians, publishers, and vendors to forge partnerships with patrons that have resulted in the development of more successful products and services for accessing e-resources. In Chapter 8, a case study by Nancy Fried Foster and David Lindahl, provides a specific example of how a library can partner with patrons in order to enhance their access to e-resources. This chapter discusses a project at the University of Rochester Libraries to analyze patron behavior in order to implement an institutional repository that will meet patrons' needs.

Part III, "Evolving Tools," consists of five chapters that discuss the new and interrelated array of tools that libraries are currently implementing in order to effectively manage and provide access to e-resources. This section begins with Chapter 9 by Charley Pennell that examines the uncertain role of the most traditional of library tools, the online catalog, in an information environment dominated by e-resources. The focus of Chapter 10, authored by Maria D. D. Collins, is a tool that is now becoming central in libraries' efforts to maintain control of their e-resources: the Electronic Resource Management (ERM) system. Highlighting ERM systems' history and discussing current initiatives, this chapter provides practical guidance for the library attempting to select and implement an ERM system. Of course, the capabilities of tools such as an ERM system are highly dependent upon the degree to which they can be integrated with a library's other tools. In Chapter 11, Mark Ellingsen provides a detailed picture of current efforts to develop standards and partnerships that will allow for the effective integration of libraries' evolving tools. In Chapter 12, authors Jeff Weddle and Jill E. Grogg provide a whirlwind tour of the diverse assortment of tools that libraries are currently utilizing in order to manage and provide access to e-resources, including A to Z e-journal lists, OpenURL link resolvers, MARC record services, and meta-search tools. The section concludes with Chapter 13, by Glen Wiley, which examines how seven academic libraries have assessed their unique characteristics in order to develop infrastructures of tools for e-resource management and access.

In Part IV, "Evolving Strategies and Workflows," five chapters are presented that explore the impact of libraries' evolving collections, partnerships, and tools on how librarians approach and carry out e-resource-related responsibilities. Elizabeth S. Burnette begins with Chapter 14, which provides direct guidance to serial workflow managers. In particular, it discusses how managers can assess their current processes and then develop and carry out an effective plan for enhancing the efficiency of these processes. Next, co-authors Jill E. Grogg and Selden Durgom Lamoureux (Chapter 15) provide readers with a discussion of the trends and initiatives that are currently reshaping one of the most challenging tasks within a library's e-resource workflows: licensing. Patrick L. Carr (Chapter 16) then builds off of Grogg and Lamoureux's chapter by discussing the specific challenges that libraries face once they have successfully acquired an e-resource, namely, ensuring that patrons' access to this e-resource is activated and maintained. Following this, Bonnie Parks (Chapter 17) shifts the discussion to the challenges that the transition to e-resources have introduced to catalogers. Finally, in the book's final chapter, Patrick L. Carr (Chapter 18) presents a case study of how five academic libraries are actually addressing the challenges of implementing workflows that are appropriate for e-resources.

Ultimately, the content of this book reflects the changing philosophies that academic libraries have adopted in order to serve patrons in an environment dominated by e-resources. Responding to patrons' expectations of quick and seamless access to Web resources, libraries realize that staying relevant requires that they demonstrate a renewed dedication to the concept of access. Indeed, focus on access over ownership is beginning to sway philosophies of collection building. Such concepts often support the just-in-time access model through such initiatives as pay-per-view and document delivery. E-resource management tools are adding an additional layer of sophistication to libraries' abilities to manage these kinds of services, make collection decisions, and efficiently meet patron demands. In today's e-resource world, patrons demand connected silos of information and access points that are consistent and require minimal instructional interference. In fact, the development focus behind many current systems initiatives and ERM tools is patron-centered, whereby patrons' needs directly impact design priorities. Consequently, the workforce required to support evolving collections and implement creative solutions for management tools is also forcing an evolution for staffing and workflow practices in libraries. Traditional practices, by necessity, are being evaluated for efficiency and evolved to address libraries' changing priorities. As librarians facilitate the transition from print to electronic with these goals in mind, their success will require an open mind toward the ultimate mission of an academic library and an understanding of how to reconceptualize serials and e-resource management to better serve this mission.

NOTE

1. T. Scott Plutchak, "What Do You Call 'Success'?" T. Scott [Web log entry on January 5, 2007], http://tscott.typepad.com/tsp/2007/01/what_do_you_cal.html (accessed January 18, 2007).

PART I:
EVOLVING COLLECTIONS

Chapter 1

Budgeting and Acquisitions

Elizabeth S. Burnette

INTRODUCTION

Throughout the academic library environment, staff are focusing much of their energy on the various issues surrounding electronic resources and digital collections. Acquiring materials in an electronic format is an evolving process. The nature of e-resources is dynamic and multifaceted, so acquiring them is complex and labor-intensive. Managing the acquisition of e-resources requires more communication and coordination than the acquisition of their print counterparts. The organizational resources used to provide access—like funding, staff expertise, collaboration, and communication—require particular coordination. The manager who oversees the acquisition of e-resources must optimize organizational resources while keeping various subprocesses in motion, much like a circus juggler. Instead of rubber balls or bowling pins, however, today's acquisitions librarian is juggling budgets, licensing, the order process, staff, and information technology. Add the publishing industry to this list, and librarians are well on their way to training for the Ringling Brothers and Barnum & Bailey Circus.

Given the often chaotic environment surrounding e-resource workflows, the budgeting and acquisitions processes for these resources have changed significantly. The objective of this chapter is to identify issues and trends that impact the management of and planning for e-resources through a discussion of budget fundamentals, expenditure tracking, and reporting. This chapter also covers acquisition topics such as working with selectors, purchase orders, subscription agents,

Managing the Transition from Print to Electronic Journals and Resources

price models, and current trends in purchasing alternatives. Through this discussion, the chapter aims to provide readers with a better understanding of the acquisitions process while illustrating the transitions that are occurring in this process.

BUDGET FUNDAMENTALS

In her book *Financial Planning for Libraries,* Ann Prentice provides an excellent philosophical description of budgeting, stating that the budget is much more than a document describing programs through the listing of items and expenditure. It is a series of conscious or implied goals with price tags; it is a contract to perform certain functions for a certain fee; and it is a mutual agreement. It is a statement of the library's expectations: what must be done and what resources are necessary to do it. Once accepted by the funding body, the budget is a precedent. New items added to the budget have a good chance of continuing. In sum, the budget is a planning tool that is basically a political document.[1] When applied to e-resources, the objective of the budget is more than setting a spending threshold or percentage. E-resource budgeting is an agreement on the financial support allocated for digital collections within a fiscal year. It is also a philosophical agreement to use the staff and information technology required to provide access. The budget is the starting point of a plan for successful access to and sound stewardship of digital collections.

In general, the acquisition of library resources starts with the availability of funding. Budgets have several aspects: (1) the financial allocation itself, (2) the process of allotting the funds for specific goals and objectives, and (3) the administration of the funds. Initially, it sounds confusing because the same words are often used to describe each of these aspects. To distinguish these aspects in this chapter, the term *budget* is used to denote the financial allocation, *budgeting* to refer to the process of allotting funds, and *tracking* to refer to allotment administration.

A budget can be defined as an allocation of financial resources to meet specified objectives. The statement of those objectives and their associated allotments are considered a blueprint for the upcoming fiscal period. As Prentice states, "the budget should be seen as a plan with dollar figures attached."[2] Library budgets are separated into two major categories, *operating* and *capital*. This chapter will only focus on operating budgets, which are funded annually to achieve the library's program for the fiscal year. The operating budget has several variations, four of which—the line-item, program, performance, and zero-base—are the most common (see Figure 1.1). The line-item budget is the traditional variety, focusing on the projections for expenditures of services. The current year serves as a basis for the next year's budget plus the estimate of increased costs and the rate of inflation. "This is not a planning budget and the library's plan is not obvious from this format" because objectives and goals are omitted.[3] However, it does display budget content and reflect spending activity, making it an effective view of the budget for the sake of managing resources.

Operating budgets can be described according to a number of particular focuses including: (1) source, (2) scope, (3) purpose, and (4) expenditure. Operating budgets that focus on the source describe funding sources; examples include government funds from tax revenue, grants, membership fees, and bequests or other gifts. Operating

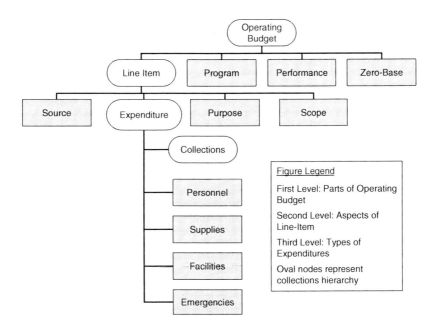

FIGURE 1.1. The Origins of the Collections Budget

budgets that focus on the scope of the budget can cover the entire library system, library locations, individual departments, or distinct collections. If the focus of the operating budget is "purpose," it will describe specific objectives or needs, like distance education. Operating budgets that focus on costs and cost management concentrate on the library's expenditures. The usefulness of the expenditure focus is the flexibility of the lines included, the ability to capture the state of the budget at a given point in time, and the ability to compare and contrast activity across reports created at different times. Some examples of expenditure budgets include collections budget, personnel costs, supplies, physical plant maintenance and emergency budgets, which are used to reallocate funds during a crisis (e.g., fire and water damage, floods, and revenue short-falls). See Figure 1.2 for examples of expenditure budgets.

The Impact of E-Resources on the Collections Budget

As a maturing format, e-resources have had a long-term impact on the collections budget. Many titles in academic collections have transitioned from print to print plus electronic to online only. So, over a

Types of Expenditure Reports

Disaster Recovery Budget
Fiscal Year

Supplies and Equipment Budget
Fiscal Year

Personnel Staffing Budget
Fiscal Year

Collections Budget
Fiscal Year

Line-Item	Allocation	Encumbrance	Expenditure	Balance
Monographs	$100,000	$10,231	$25,442	$64,327
Subscriptions	$800,000	$45,889	$250,242	$503,869
E-Resources	$500,000	$79,203	$198,222	$222,575
Total	$1,400,000	$135,323	$473,906	$790,771

Page 1   1/15/2008

FIGURE 1.2. Examples of Expenditure Budgets

short period of time the funds required for e-resources have grown exponentially in response to the number of titles that have shifted from the print format or added an electronic component. For many libraries, e-resources were first considered as part of the serials subscriptions line-item within the collections budget. The e-resource format is now significant enough for a separate line-item to track activity in greater detail over time. However, it will continue to affect the print serials line-item as more titles transition across formats. Funds must be transferred between the print serials and the e-resources line-items. Furthermore, the availability of titles in both formats simultaneously has made budgeting by format less distinct.

With titles moving away entirely from the print format, selectors in some libraries are now beginning to refuse print and select electronic-only for targeted areas. This trend is slowly changing the composition of library collections in addition to the collections' value. Physical materials are giving way to virtual resources. Administrators are noticing that, as expenditures for print decrease, the capital asset within libraries will not increase at past rates and, over time, will ultimately stall and decline. Johnson notes in *Fundamentals of Collection Development and Management* that, unlike their electronic counterparts, well-maintained, physical materials can appreciate over time.[4] Even if it takes ten to twenty years to manifest, this evolution will have lasting repercussions on the status of the library within the parent organization.

In the meantime, the nature of e-resources both complements and challenges the libraries' budgets. E-resources complement the libraries' budgets by alleviating costs. As Johnson states:

> Some library costs associated with print products disappear if the library moves away from print subscriptions. Among these are staffing costs to [receive] issues and claim missing issues, circulate items, manage shelving, order replacements if needed and bind into volumes—along with the cost of binding and repair of physical volumes.[5]

However, with these cost variations, it does become difficult to compare subscription costs across formats. Nevertheless, Johnson points to research that Carol Hansen Montgomery carried out at Drexel University which indicates that, on a cost per-use basis, e-journals are a more cost-effective choice for libraries than print journals. The primary reason for this finding is that e-journals require no physical shelf space; storage is a major expense of low-use bound journals.[6] The rate of inflation will also impact library budget planning. Ultimately, as the digital format matures, "collections librarians should articulate both cost savings and the less tangible value accruing in improved access to materials and user satisfaction."[7]

In contrast, e-resources challenge the libraries' budget by increasing other costs. There is a long-term impact from the direct costs—including inflationary increases—to renew the resource. In addition, the overall budget is impacted by supplemental costs incurred via expenses paid to acquire the title, maintain the subscription, upgrade equipment, educate users, and negotiate and manage licenses. These costs redirect funds for other purchases and processes as libraries decide how to fund the various costs associated with e-resources, such as backfiles, servicing, managing, and accessing electronic information. Other possible costs affecting the budget include hardware, software, and routine information technology staff time and support.[8] Understanding how e-resource pricing structures affect the budget creates a helpful framework for the decisions managers must make. Budgeting for direct costs is only a partial solution. Monitoring the supplemental expenses and indirect costs is also required to steward the shrinking collections' dollars.

Tracking E-Resources

An important part of budgeting for e-resources is the ability to link the allotment for the format (the e-resource line-item) to the acquisition of the e-resource title or package. This link will provide the foundation for effective tracking during the fiscal cycle. The link also connects the staff members designated to steward the funds to the digital materials acquired. As Johnson states, "ideally, responsibility for managing funds should be consistent with policies that assign responsibility for selection and collection management decisions."[9] So, precisely how can libraries efficiently track e-resources throughout the acquisitions cycle? What links the allotment to the title? Regardless of the mechanism—integrated library system or in-house database—tracking the e-resources allotment is necessary for effective budget management. This becomes more important as new titles move to the digital format and prices escalate during access renewal. Several fund models can be used for e-resource tracking systems:

1. A single, central fund line can be established to track purchases by format only, all e-resources distinct from print serials. An example is to create a digital or an electronic fund to store the budget allotment for the fiscal cycle. The advantage is simplicity for selectors and staff. The disadvantage is little granularity with fund statistics; a separate system for tracking expense detail is required.

2. At the other end of the spectrum, distinct subject funds can be established to identify broad subject classifications and/or core research areas that need support. An example is to identify important subjects and design a fund for each. The advantage is the detail available for expense tracking and reporting to departments. The disadvantage is the increased complexity of managing funds by selectors and staff.

3. Combining the models described in previous text can create a central fund for general items plus subject funds for areas of research impact.[10] This model incorporates the cost benefits of both—the ease of the first and the detail of the second. A hybrid of these models can be configured for more complex tracking with a central fund to track the format plus subject funds to track the noteworthy packages, for example, consortial purchases. The labor involved is proportional to the quantity of noteworthy packages being monitored.

Also, tracking e-resources contributes to the staff's understanding of the budget purpose, how it is planned, and how the collections budget is dispersed. Furthermore, with an understanding of e-resource data standards, staff can create appropriate nomenclature and codes to track e-resources using the mechanism at hand. The use of fund sub-accounts in the financial database or Integrated Library System (ILS) will link this data together. When using an ILS to track funding, responsible staff should understand the policies and configuration of the fund module in order to maximize the potential of this system and facilitate accurate reporting.

To successfully implement tracking, it is necessary to document and communicate the fund names, identification codes, and allotments to all selectors for use during the spending cycle. A general understanding of how tracking is achieved makes it easier for staff to create or comply with procedures throughout the acquisitions cycle. Continuing education is required to address both attrition and/or adjustments to the budget or technology used. Implementing any tracking requires that each selector commit to consistent use of data required for the tracking mechanism. Acquisitions staff can use this information to establish and maintain the budget tracking and reporting tool(s) in addition to monitoring the actual expenses incurred.

When it comes to tracking e-resources, several challenges make the process less clear than that of its print counterparts. The dynamic nature of the digital format and its interdependence with print makes budgeting more difficult. How do individual funds support titles that are acquired in multiple formats? As offerings from publishers change, how should selectors plan requests for allotments? What is the impact of cancellations on funds? Tracking current expenditure activity contributes to the planning discussion required to sort through these issues. The ability to report the outcomes of this planning and spending is a key element in this dialogue.

Reporting E-Resources

E-resources are difficult to plan for because selectors need to both respond to the publishers' offerings and proactively select formats to meet research or users' needs. It is challenging to anticipate which titles will be acquired or renewed in an electronic-only format, so reports are an essential tool for e-resource budget management. A battery of reports is needed to fully address the diverse needs of managers and stakeholders who plan and monitor spending throughout the fiscal year. The expenditure report is one tool that can aid acquisitions in monitoring line-items descriptive of current spending trends (categories may include e-journals, e-resources, e-books, etc.). Helpful fields to display in such a report include allotments, encumbrances on outstanding purchase orders, expenditures paid to date, and balances remaining. Monitoring these categories becomes especially important at the end of the fiscal year when acquisitions staff work to exhaust fund balances, monitor open orders and unexecuted licenses, and follow up on unreceived invoices. The expenditures and balances accomplish several objectives: (1) display over-/under-spent lines, (2) indicate when the allotment is insufficient, (3) highlight when unusual spending occurred in a line-item, and (4) point out when new trends can be better tracked using new allotments.[11]

The expenditure report responds to inflationary factors to provide continuity when used consistently across years or library locations. However, because the report does not reveal an actual plan with goals and objectives, it should be considered only one of the documents needed for comprehensive budget planning. Generally speaking, the expenditure report for the current fiscal year can serve as a basis for the subsequent year's budget plus the estimate of increased costs for the rate of inflation.[12] Nevertheless, e-resources are somewhat difficult to plan around because their overall packaging and pricing can change significantly over a brief period of time. Therefore, expenditures for a given year's e-resources may have less bearing on the needs for subsequent years; however, they do serve as a starting point for a budget plan.

Planning for new e-titles—available for the first time in digital form—is another complicated aspect of the collections budget. Traditionally, libraries created line-items under historical order types: monographs and serials. Later, formats became more diverse within order types and expenditure reports contained lines including microforms and newspapers. E-books are often purchased in aggregated packages, blurring the distinction between the firm and continuing/serial expenditures. Indeed, these expenditures could qualify for either the monographs or the serials portion of the report. To align reporting with current e-resource packages, reports can be revised to distinguish monographs, serials, and electronic order types. This change would facilitate the monitoring of changes in both print and e-resource purchasing. After the recent publishing surge of print-with-electronic offerings, more publishers have generated enough revenue to cease print and offer electronic-only titles. Libraries have become more comfortable with electronic-only access, and preservation issues are being addressed via licensing and by products like JSTOR, Portico, and Lots of Copies Keep Safe (LOCKSS). Currently, many libraries are either cancelling print subscriptions or investigating this option. Over time, expenditure reports that track e-resources will help to document this transition and facilitate trend analysis as well as continued planning.

With both a budget allotment and a tracking model, managers who design budget reports should meet with staff who monitor the budget process (e.g., administrators and selectors) to confirm the overall interpretation of what constitutes the e-resource line-items and how they should be presented. The presentation of e-resource budget data on reports can be organized around the following purposes: planning, statistics, expenditure tracking, collection maintenance, and oversight. Communication and collaboration will ensure that all stakeholders agree on the standards and interpretation before a report is distributed. A thorough report will allow staff to report, track, and administer e-resource funds throughout one and across many fiscal years. Collaboration on the logistics of the reporting process will also allow staff to determine who should be involved in monitoring the budget, the frequency needed for reporting, and how the data will be shared (electronic posting, e-mail, or print).

ACQUIRING E-RESOURCES

Over the past fifteen years, advances in information technology and changes in the publishing industry have transformed the face of library collections and the work performed in acquisitions departments. Prior to e-resources, workflows for the acquisition of print serials were typically performed in an isolated process that required only basic communication across the boundaries of technical services. E-resources ushered in an evolution that required acquisitions staff to communicate consistently outside of technical services. They now collaborate routinely throughout the library with staff in administration, reference, public services, and information technology regarding such issues as business terms, trial access, pending license agreements, access parameters, and renewal terms. These broadened communications are evidence supporting Bosch, Promis, and Sugnet's statement that "traditional boundaries between technical services, collection development and management, information technology, and public services [have] become blurred and indistinct when dealing with electronic information acquisitions."[13] Multiple departments have a hand in the acquisitions process and must work in concert to interact with content sources, establish access, and serve users.

Within the acquisitions department, the line has also blurred between monographic and serials processing units. E-resources have defied traditional processes and created a new collaboration between staff accustomed to working independently on specific formats. Unlike e-resources, purchasing print books and serials is relatively straightforward and somewhat linear (see Figure 1.3).

However, the process of acquiring e-resources is not linear; the format has a substantial impact on the process. Due to their varied nature and virtual format, acquiring e-resources is more complex, as Figure 1.4 demonstrates.

E-resources can be available from several sources, use different software platforms, and have a variety of pricing models.[14] In addition, acquiring e-resources is a longer process with potentially many steps and requires more attention from staff to monitor activity and flow. An analysis of some of the steps included in Figure 1.4 will follow with a description of how e-resources have impacted the acquisitions process.

WORKING WITH SELECTORS

During the era of print, selectors could provide a title to acquisitions staff adept at finding materials from publishers and vendors. Only in the instance of obscure, foreign, or rare materials would a selector indicate the best or sole source for an item and its price. With e-resources, selectors are more consistently communicating directly with the source on purchase terms. In addition, Gregory notes that

> the traditional primary skills of a selector now must be augmented by knowledge of the technical aspects of the materials, consideration of the copyright and licensing issues that may be implicated, as well as familiarity with the various bundles of electronic materials.[15]

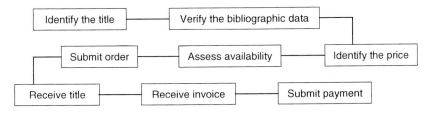

FIGURE 1.3. Basic Acquisitions Process for Print Serials and Books

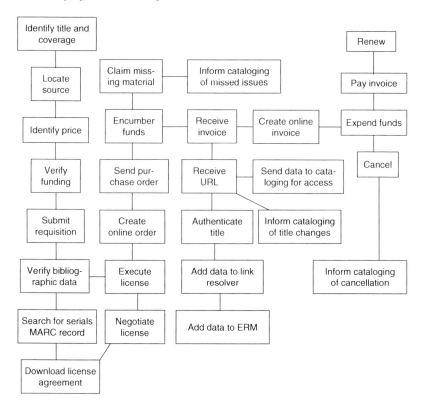

FIGURE 1.4. Potential Steps in Acquisitions Processes for E-Resources

Also, the evaluation of e-resources can be much more complex than print. For example, in *Selecting and Managing Electronic Resources,* Gregory provides a handy checklist of selection criteria not normally considered for print, including the cost of outright ownership, the requirement of a license, title-/article-level purchase options, and the availability through a consortium.[16] Due to the increase in the complexity of such selection criteria for purchasing e-resources, an even greater degree of collaboration is required between selectors and acquisitions staff to determine the details necessary to initiate and carry out the order process.

Other complications that have extended the role of the selector exist. Today, the economy, annual fluxes in library funds, and inflation impact the collections budget on a continuing basis. Over time, these factors, combined with the proliferation of worldwide publishing, e-resources, and users' expectations, have resulted in an evolved process required for the selection of appropriate resources for the collection. According to Johnson, selectors must now "balance more systematically what is acquired locally with what is accessed remotely and borrowed from other libraries."[17] Also, the emergence of aggregator packages has returned specific title selection back to the source. In these instances, the selector role has been relegated to selection on the package-level rather than the content-level.[18] This is further evidence that the work of selectors is evolving. The result of this evolution is that selectors no longer simply pick a title and cancel subscriptions. To manage e-resource selection, selectors need to collaborate with both systems and acquisitions librarians to compare options and assess IT specifications. Analysis data, user evaluations, and assessments are also useful tools that can assist in guiding this process.[19] Finally, as selectors engage in a dialogue with library staff to communicate issues concerning various selection criteria in addition to philosophies and plans for building the collection, they are increasingly assisting the entire e-resource life cycle.

ORDER PROCESS

Acquiring e-resources has the same phases as other formats: selection, order, payment, access, and storage. In respect to the order phase of e-resources in particular, numerous factors must be communicated between the library and the vendor. Occasionally, the license negotiated for purchase will take the place of the actual order for communicating these specific items. Nonetheless, a paper trail is still required to track the encumbrance and document the payment.[20] According to Bosch et al.: in addition to the standard information on serials purchase orders (e.g., starting volume and coverage), orders for e-resources typically require that the library provide the vendor this additional information:

- Type of access (Web access, etc.)
- Institutional IP addresses, including proxy servers
- Technical contact name, address, phone and fax numbers, e-mail address

- Product version
- Expected start date
- Number of simultaneous users (if applicable)
- Bill-to and ship-to addresses, as well as addresses for the technical contact and a contact for licensing issues
- Brand of computer for software (e.g., Apple)
- Complete description of the package being ordered if purchasing bundled hardware and software, including
  —RAM capacity
  —CD-ROM, DVD
  —Monitor
  —Software to be included (Windows, etc.)
  —Keyboard style, mouse, etc.[21]

Another requirement often needed for an e-resource purchase is the license agreement. The negotiation process to finalize a license can frequently extend the timeline for fulfilling an e-resource order. As Bosch et al. state, "the purchase order is a formality; the signing of the license is considered the point of purchase."[22] Presently, there are few mechanisms in e-resource acquisitions that can expedite fulfillment of a licensed product, like the rush order or pro forma invoice for print publishing. Also, few publishers will set up subscription access prior to executing a license agreement.[23] Doing so can complicate the payment terms for the product.

Another challenge particular to acquiring e-resources occurs when the purchase contains both one-time and continuing expenses. Traditionally, monographic publications have been restricted to firm orders and serial publications to continuing orders. Some acquisitions departments with sufficient personnel have even built separate units around this distinction. Often, however, the distinction between the monographic and serial order processes for e-resources may become blurred, and specialized staff may find themselves operating in unfamiliar territory. For example, e-books that are sold with multiyear access fees require a continuing commitment from the collections budget and extended tracking by staff, which may be composed of personnel who traditionally work with monographs. Also, e-journals are often sold with access to backfiles, which may require a firm order. Effective communication regarding business terms is important in these instances. Product packaging, along with fees for perpetual access and price models, should be clearly investigated during the selection and purchasing process. Guidelines for funding these variables should be established and the parties involved need to remember Bosch et al.'s statement that "the anticipated future cost of the product would be based on payments for the serial expense [and] the expenditure for the backfile should not be used in cost projections."[24] Consideration of these concerns in addition to consistent practices for placing orders for both monographic and serial aspects of e-resources will contribute to the availability of the data needed later for audit controls, accreditation, and planning.

Acquisitions staff should be aware of additional concerns during the order process for e-resources as well. One such concern is prepayments, which should be handled with care, especially when staff are learning the numerous subprocesses needed to acquire e-resources. Also, internal mechanisms and communication loops need to be established to alert staff when access is provided and enabled so that access points can be established. After the order is established, separate procedures may be needed to track claims for e-resources. As more of the collection shifts to electronic-only access, claims will evolve into a process to track errant e-resource access. E-journals are trickier to claim than print, as the timelines of new issues makes these resources more difficult to track.[25] Fortunately, Electronic Resources Management (ERM) systems help to address many of these concerns by connecting the various ends of the workflow and establishing a system of alerts to facilitate follow-up of these tasks.

Finally, the cancellation process for e-resource orders, which is often very similar to that of print serial orders, is also important and should be taken into consideration. For both print and electronic serials, there are several reasons why a library may choose to cancel unfulfilled orders after generating claims to the vendor:

- when receipt is unlikely, priorities change, or an alternative source becomes available such as gifts or complimentary shipments;
- to address concerns about expense or lack of use;
- to release encumbered funds to facilitate end-of-year processes.[26]

Of course, the terms of the license agreement specifying cancellation allowances and/or conditions is one distinct difference between print and electronic formats. Library staff should be aware of any cancellation language and stipulations found in license agreements. Bosch et al., emphasize this fact, stating that "an electronic information product license should specifically define the terms for cancellation. If it does not, procedures and routines should be established for cancelling orders with the vendor or publisher."[27] When a library cancels a title, previous agreements regarding perpetual access and archival rights should be carefully considered, and acquisitions staff will need to make sufficient arrangements to ensure this access. In respect to payment terms after cancellation, staff may need to negotiate a credit or refund with the vendor. Some publishers require the destruction or removal of data from local systems. When physical materials are involved in the cancellation, libraries may have to return items to the publisher accompanied by written communication of the cancellation. In the absence of a protocol to cancel e-resources, library staff should at least note the status, selector or person making the cancellation decision, and effective date for the cancellation in the order record. The order record can also be used to note when the vendor communication for cancellation took place.[28]

SUBSCRIPTION AGENTS AND THE ACQUISITIONS PROCESS

Subscription agents are effective intermediaries in the acquisitions process for libraries when it is not possible or advantageous to purchase directly from the publisher. They provide services that add value to their relationship with libraries such as electronic invoices for local systems and title-specific or itemized payment information for budget tracking. Many publishers lack the capacity for both. These agents also manage title lists, renewals, and invoicing for libraries and publishers. In addition, proprietary interfaces are often provided to customers to assist in the management of the acquisitions information or processes such as format options, registration instructions, and licensing. Publishers, in contrast, have a clientele far too diversified to make this kind of automation affordable.[29] Subscription agents also consolidate payments, claims, and proof of payment and can often provide detailed invoice information for local systems. All this becomes urgent when library staff are troubleshooting e-resources access problems. Subscription and payment history facilitate the critical problem-solving process required to reestablish access.

To assist the budget process and assessment efforts of libraries, agents can also provide collection development reports, pricing, and projections. Mediation between publishers and customers when concerns grow from the customer vantage point is another service subscription agents provide.[30] Despite fees for various services, subscription agents can often more cheaply support single-subscription electronic purchases than a library can on its own. Publishers may offer the most cost-effective support for large, subscription-based packages. The costs and benefits of working with an agent or the publisher should be assessed before buying to determine if value-added services are more important than service charges.[31]

PRICE MODELS FOR E-RESOURCES

Another factor influencing the acquisitions and budgeting process for e-resources is the variation of dynamic price models offered. Publishers create packages and base pricing on factors like company direction, the size and type of the customer organization, and simultaneous users. Discounts are sometimes offered to offset the expense for multiple copies, but are not standardized due to the variety of products, vendors, and publishers.[32] Current models include the following:

1. *Subscription-based* pricing is designed to move publishers from print to electronic publishing while protecting revenue streams. An advantage for libraries is the option to purchase the subscription in two formats. This category includes three types:
   a. *Electronic-only subscription* encourages libraries to switch from print to an electronic format. Costs are saved for the publisher who will no longer produce and ship print copies and the library which avoids binding and storage. The risks of electronic-only subscription tie into the lack of permanence for archives. New initiatives are contributing to the permanence of the digital format via products like Portico and LOCKSS.

    b. *Electronic-plus-paper with deep discount* reduces costs accrued when both formats are needed. The advantage is the affordability of the two formats. This model remains popular because of the access to the owned print that is archived locally. There are costs for the subscription and additional costs for the "extra" electronic access. Or, a basic cost for the online copy and additional costs for the paper.

    c. *Additional journals package* requires the maintenance of current subscriptions in exchange for access to all the electronic titles available in that given package. The advantage for libraries is expanded title access, while the disadvantage is access to titles a library may not need. This model, popular with consortial deals, protects the publisher's bottom line by extending the length of the contract and cancellation restrictions. A variation is the "big deal" or the whole-package model in which current subscriptions are maintained in exchange for access to all of a publisher's electronic titles. There are pros and cons and long-range effects with these deals; with expanded access, the library becomes locked into these packages and loses flexibility in the collections budget. If the additional titles are marginal, this model may not be worth the loss of purchasing power for the library.

2. *Access-based* pricing considers the number of simultaneous users or ports purchased. The disadvantage for libraries is that access is limited to the agreed upon number of users at one time. Beyond the maximum, users are turned away, but the library purchases the access it needs, which can be more affordable than unlimited access. Usage reports are then utilized to gauge how much access is needed over time and when adjustments should be made.

3. *Site license* pricing provides access to an unlimited number of users based upon the number of full-time equivalent students. This is effective for high usage products and often, the more access purchased the cheaper the cost per user. This model can have either a fixed price structure or a price relative to the potential number of users.

4. *Pay-per-use* is associated with full-text products. The advantage is affordability; it provides libraries with low volume access to

an expensive product. However, publishers require sophisticated billing and delivery mechanisms to manage this model.

5. *Subscription with pay-per-use core titles* is an alternative that provides access to subscription at a subscription price with additional content on a pay-per-use basis. This can be very cost-effective for libraries. High-use content of great expense comes via a subscription while titles that are not cost-effective due to questionable use per costs are available as needed.

After the appropriate model for purchasing a given resource has been selected and negotiated, the length of time for this process requires that staff check to ensure that aging price quotes will be honored during the procurement phase of the purchase cycle. Bosch et al., go on to comment that "the electronic publishing industry is changing rapidly, and the pricing structures are shifting accordingly."[33] Libraries can capitalize on this dynamic. The fact that pricing structures are evolving opens an opportunity to explore creative models during negotiations such as offering a specific dollar amount to see what it will afford, a simple bottom-line price offer, or custom packaging to align contents with existing collections or user demands.[34]

Alternatives to Traditional Purchasing

The combination of shrinking collections budgets and the rising cost of serials and e-resources have made it difficult for libraries to provide users with additional access to e-resources. Meanwhile, technology continues to advance exponentially and is impacting publishing, content delivery, and users' expectations. This scenario has sparked two alternatives to tradition material ownership models: (1) on-demand access and (2) consortial purchasing.

According to Johnson, "since the 1980s the profession has given much attention to what is known as the access versus ownership debate. This phrase describes the choice between deciding what to own locally and what to access remotely."[35] Ownership is self-explanatory, but access addresses temporary availability, covering leases and licenses for electronic transmission, interlibrary loan services, and document delivery services. Access versus ownership is paired with a just-in-time concept, a business phrase describing the reduction of buffer inventories in favor of the synchronized movement of materials through production immediately prior to their need and use. Access aligns philosophically with this just-in-time concept, while ownership aligns with the just-in-case philosophy traditional to libraries. Johnson provides further analysis of these two competing concepts for access, stating that:

> A library can be said to follow a just-in-time approach when it acquires either through purchase or loan materials its users need when they need them and does not invest all or large portions of its materials budget in acquiring collections just-in-case users will need them at some future time.[36]

Just-in-time or on-demand access to materials is the first alternative to traditional purchasing.

User aptitude and demand have kept pace with technology, and the result has been greater pressure upon libraries to meet the needs of a more diverse population with shrinking financial resources. As serials studies have indicated, savings can be realized when libraries strive for ownership of high-use materials and access to low-use or expensive materials. Document delivery and interlibrary loan services are a means to on-demand access. They continue to serve a vital role in non-acquisition services; as Johnson states, "interlibrary borrowing is an integral element of collection development for all libraries, not an ancillary option."[37] Publishers have responded to these issues and now offer publications and pricing for on-demand acquisition of books and e-resources. Also, publishers will now establish institution and individual accounts for per-title access to the journals in their catalog. Another benefit to on-demand access purchasing is that libraries can avoid the pitfalls of subscribing to bundled titles that limit future title cancellations and diminish budget control.[38]

The second alternative to traditional acquisitions is consortial purchasing of e-resources. Library consortia are defined as associations of libraries that strategize to share funding, materials, technical expertise, and risks. According to Bosch et al., "most academic and public libraries in the United States are affiliated with one or more consortia."[39] The benefit is in the consortium's ability to gather multiple buyers that can combine purchases into one transaction for the publisher. The local needs of a library should influence its affiliation with other libraries and the benefit it stands to gain. Bosch et al., indicate that the following factors influence the partners that a library may enter into on a consortial relationship with

- types of member institutions and the legal requirements of governing bodies;
- resources available—both financial as well as personnel;
- environment, including current corporate cultures, technical expertise, and technical infrastructure;
- expectations and scope;
- time frames, including renewal cycles.[40]

The issue of the storage of print materials has created obstacles for consortia choosing to reduce redundancy between collections but failing to negotiate the stewardship of core materials needed by the user populations they serve. In contrast, the electronic format was ripe to capitalize on collaborative purchasing for group savings because these materials did not require the commitment of shelf space and storage.[41] For many institutions, the issue of expense rendered many important e-resources out of range for serials budgets straining under the weight of print renewals and electronic endeavors. However, these same institutions have been able to take advantage of consortial purchasing opportunities due to publisher discounts not made available to individual institutions. Since the publisher is often the sole source for an electronic resource, it can discount based on volume, concurrent ownership of print or other formats, or prepublication status.[42]

Libraries are often members of multiple consortia of which one may be highly organized with staff, office space, budget, and nonprofit corporation status. On the opposite end of the spectrum are less-structured groups with members sharing the work and moving by consensus. This gives libraries the option to affiliate simply for a single product or for multiple options to purchase the same product with several consortia.[43] Usually, the library is a member of an Online Computer Library Center (OCLC) regional consortium and a statewide consortium.[44] When the consortium has multiple missions, it may be less successful at acquiring e-resources and purchasing products that are broad in scope. Consortia vary one from another, but each possesses some common traits, including:

- value-added services to members;
- reduced costs or avoided costs;

- effective organization and leadership;
- effective lateral and vertical communication;
- sustainable organization;
- adequate resources (including funding as well as personnel);
- flexibility of the organization's response to a changing market and needs.[45]

Presently, statewide consortia have provided the best alternative to the individual purchase of expensive e-resources. E-resource acquisition is now a major purpose for these statewide consortia. During license negotiations, their collective clout can persuade publishers to modify their positions.[46]

Consortial purchasing has become prevalent for several reasons: (1) format, (2) shared costs, (3) easier licensing, and (4) cheaper overhead. The ease of sharing e-resources due to the electronic format defuses the argument over who keeps what and where it is housed. This was often a sticking point with print resources and sometimes a deal breaker. By sharing expenses, each library pays less and gets more by pooling their financial resources.[47] This concept ties into the willingness of publishers to be flexible because of the revenue available from the consortium. Publishers will often improve these deals by raising the threshold on specific license terms for consortia that are not available to the individual, like interlibrary loan and document delivery options. Publishers hope that more parties will become interested in joining the agreement. Consortium licensing staff also gain valuable experience as negotiators, impacting future arrangements. With purchase costs and license efforts spread over multiple institutions, it is cheaper for each member to get through the acquisitions cycle.[48]

Consortial purchasing is more time-consuming than purchasing direct because several organizations need to agree. Generating a consensus is time-consuming and not expedient for titles needed in a short time frame. It can be more beneficial to pursue consortial agreements for such products well in advance of their next renewal. Some licenses will allow nonparticipating members to join the purchase at future dates. Some purchases provide open-ended arrangements to members who may purchase the product at a future date.[49] The improved terms of the purchases extended to consortia compensate for the time needed to see the purchase through. The amount of time needed and procedures used vary within a consortium and also vary according to the product under consideration.[50]

CONCLUSION

The evolution of e-resources has had a widespread impact on library budgeting and acquisitions. Understanding this impact creates a framework for managing digital collections. The e-resource line(s) within the collections budget is both a plan and a goal; it is also a fiduciary agreement to use funds, staff, and information technology to provide access for users. Titles in this format impact funds allotted for print serials and complement the budget by alleviating various costs associated with print. At the same time, e-resources challenge the budget by adding supplemental costs via expenses paid to acquire the title, maintain the subscription, upgrade equipment, educate users, and negotiate and manage license agreements.

Managers need to monitor the dynamic e-resource price models and their effect on purchasing power, budget administration, and work procedures. In addition to budgeting, tracking, and reporting the e-resources budget are necessary functions for responsible management of the collection. Whether by ILS or in-house database, the appropriate fund model can provide the convenience and/or granularity needed for planning, statistics, expenditure tracking, collection maintenance, and oversight.

E-resources have ushered in an evolution that requires acquisitions staff to consistently communicate both within and outside technical services. More departments have a hand in the acquisitions process and must work in concert to negotiate beneficial purchasing terms for e-resources and establish effective access points that meet the expectations of users. When available, a subscription agent can effectively facilitate the process to acquire e-resources by offering value-added services.

Pricing models make product comparisons extremely difficult because there are no standards. Model changes can prove advantageous to libraries interested in pursuing creative options not already on the menu. Cost combined with other factors has spurred both on-demand access and consortial purchasing alternatives to traditional ownership models. In addition, consortial purchasing of e-resources creates discount opportunities for libraries that can collaborate together and pool resources to meet users' needs. Both favor the users and boost the impact of the shrinking collections budget.

Acquiring materials in digital form is both complex and labor-intensive. Efforts to manage the communication and coordination needed to achieve access to e-resources will help forge a necessary dialogue between librarians and staff resulting in more efficient workflows and acquisitions processes. Furthermore, an awareness of the impact that e-resources have on acquisitions will help librarians juggle competing demands for effective e-resource management.

NOTES

1. Ann E. Prentice, *Financial Planning for Libraries,* 2nd ed. (Lanham, MD and London: The Scarecrow Press, Inc., 1996), 23.

2. Ibid., 19.

3. Ibid., 36-7.

4. Peggy Johnson, *Fundamentals of Collection Development and Management* (Chicago, IL: American Libraries Association, 2004), 217.

5. Ibid.

6. Ibid.

7. Ibid.

8. Ibid.

9. Ibid., 218.

10. Ibid.

11. Prentice, *Financial Planning for Libraries,* 85.

12. Ibid., 37.

13. Stephen Bosch, Patricia A. Promis, and Chris Sugnet, *Guide to Licensing and Acquiring Electronic Information* (Lanham, MD: Scarecrow Press, Inc., 2005), 33.

14. Vicki L. Gregory, *Selecting and Managing Electronic Resources* (New York: Neal-Schuman Publishers, Inc., 2000), 35.

15. Gregory, *Selecting and Managing Electronic Resources,* 38.

16. Ibid., 40-1.

17. Johnson, *Fundamentals of Collection Development and Management,* 23.

18. Brian Quinn, "The Impact of Aggregator Packages on Collection Management," *Collection Management* 25(2001): 58.

19. Johnson, *Fundamentals of Collection Development and Management,* 23.

20. Bosch et al., *Guide to Licensing and Acquiring Electronic Information,* 39.

21. Ibid., 40.

22. Ibid., 41.

23. Ibid.

24. Ibid.

25. Ibid.
26. Ibid., 44.
27. Ibid., 43.
28. Ibid., 43-4.
29. Ibid., 49.
30. Ibid.
31. Ibid., 50.
32. Ibid., 46.
33. Ibid., 48.
34. Ibid.
35. Johnson, *Fundamentals of Collection Development and Management*, 255.
36. Ibid., 256.
37. Ibid., 255.
38. Gregory, *Selecting and Managing Electronic Resources*, 39.
39. Bosch et al., *Guide to Licensing and Acquiring Electronic Information*, 64.
40. Ibid., 66.
41. Gregory, *Selecting and Managing Electronic Resources*, 38.
42. Bosch et al., *Guide to Licensing and Acquiring Electronic Information*, 48.
43. Ibid., 66.
44. Ibid., 64.
45. Ibid.
46. Ibid.
47. Ibid., 64.
48. Ibid., 65.
49. Ibid., 67.
50. Ibid., 67.

Chapter 2

Criteria for Selecting
and Evaluating E-Resources

Christine L. Ferguson

INTRODUCTION

The number and variety of e-resources available to libraries have increased exponentially in past years. Patrons expect to find resources online and available to them at any hour of the day and from any location. Often, the challenge librarians face is not whether to select the print version or the electronic version of a resource, but rather, to select the most appropriate e-resource. There are many types of e-resources available to libraries, including e-journals, abstracting and indexing (A & I) databases, full-text databases, e-books, and free Internet resources, to name a few. In many cases, the lines between these resources are blurred. A & I databases may incorporate full-text content or provide direct links to the full text in other resources. E-journal collections, from a single publisher such as the American Chemical Society Web Editions or from an aggregator like JSTOR, can be searched for citations in the same way an A & I database can. A single journal title may be available in several different e-resources, with different dates of coverage in each resource. This cross-purposing of content can make it even more difficult to come to a clear-cut decision as to whether or not to acquire or to provide access to a resource.

Print subscriptions are usually evaluated on the basis of two very broad criteria: content and pricing. In other words, what is included in the subscription, and how much does it cost? Other factors, such as the reputation of the publication or where it is indexed, may also be taken under consideration. In some ways, the selection and acquisition of e-resources is similar in nature to the selection of print resources. Factors such as coverage, authority, relevance, audience, and price are examined, and in the selection process librarians often make use of resources that review various titles and products. However, the selection of e-resources covers a broader spectrum of criteria, including factors such as accessibility and design. For example, while you may not typically evaluate a journal based upon the design of its cover and the organization of the articles within it, design takes on new significance when evaluating an e-resource. A database's interface design and organization of content will directly impact if and how users will make use of the resource.

Whether you are having difficulty selecting the best platform for an e-resource or trying to choose the most appropriate resource for a specific subject, having a clearly defined set of selection criteria can help guide an organization through the evaluation and selection process. This chapter will outline and discuss some of the criteria specific to the selection of e-resources. Rather than providing an extensive review of the literature, the focus will be on providing direct guidance and advice to librarians, based upon firsthand experience with e-resource selection. As each library has unique needs, the criteria mentioned here are not intended to be comprehensive, but rather to provide a firm base upon which a more complete selection policy can be built.

PRELIMINARY QUESTIONS

When considering the acquisition of an e-resource, it is helpful to first ask a few preliminary questions. These questions will help guide the selection process and enable you to determine which selection criteria will ultimately carry the most weight when making a final decision. First, will this e-resource serve as a replacement for or a supplement to an existing resource? If the e-resource being considered is intended to be a replacement for an existing resource, it may be necessary to be more stringent in the evaluation. However, if the resource is meant to be a supplement to other resources, it may be possible to be more lenient. Second, before beginning the evaluation, consider how users will actually utilize the resource. What is the primary need this resource is supposed to fill? A resource that is meant to be a reference resource for quick consultation will have different criteria than one that is meant to provide more in-depth coverage.

Once the basic parameters for the evaluation of the resource are defined, you are ready to begin the actual evaluation. The criteria for the selection of e-resources can be divided into five broad areas: content, design and usability, accessibility, licensing, and pricing.

Content

Content selection criteria assess the quality and quantity of the materials covered in the e-resource. Exactly which criteria are used to evaluate the content of an e-resource will vary depending upon the type of resource being considered. If the resource consists primarily of full-text content, issues such as the file format of the full text and whether or not graphics are included will be of importance. If the resource is an indexing database, factors such as whether or not abstracts are included may carry more weight. It may be helpful to begin the evaluation process by determining what kind of resource you are considering and creating a list of criteria specific to that resource.

When evaluating the content of a resource, one of the first steps is to assess how many titles and/or item(s) are included in the resource. How many journals or articles are indexed? How many are available in full text? If you are evaluating an electronic encyclopedia, how many entries are included? Once you have determined how much content is available, compare that to your existing subscriptions—print, electronic, and microform—to determine duplication. When evaluating duplication, it is important to consider all formats, not just electronic. For example, while you may find that a full-text database has 25 percent duplication with your existing e-resource subscriptions, further investigation may reveal that 50 percent of the content is duplicated when you consider your print and microform subscriptions. You will need to determine what your library considers an acceptable level of duplication. It may be the case that electronic access to a particular resource is so important to your institution that duplication with print is not a factor in your decision.

Evaluating the extent of the full-text content of e-resources adds an additional level of complexity to the evaluation process. It is not enough to determine that a resource contains full text. The amount of full text and the way it is made available plays a role in the evaluation. Consider the amount of full text available electronically in comparison to the print version of the same resource. Is there cover-to-cover access? Is nontextual content, such as figures, tables, and graphics, included? The file format of the full text is also important. Is the content available as PDF, HTML, both, or some other file format altogether? Users that want to be able to cut and paste content into documents may prefer HTML, while those that are interested in viewing or printing an item as it appeared in its original publication, including image content, may prefer the PDF file format.

The rate at which content is added and updated is another important factor that can impact the decision whether to acquire an e-resource. Is there a time lag (embargo) from the time the content appears in print to when it is available electronically? If you are looking to replace an existing print subscription with the electronic version, lengthy embargoes may prevent you from doing so. If there is no embargo on the full text, how quickly is new content added? Oftentimes, e-journal content is made available online before the print issues of the journal are mailed. It is also essential to consider how long users are willing to wait for access to the electronic content. How frequently is the resource updated? What is an acceptable frequency for updates? For example, the electronic version of an encyclopedia set may be updated only once a year, whereas an e-journal may be updated much more frequently (e.g., monthly or even weekly).

Many of the content criteria described in previous text are quantitative. A more nebulous content criterion is the nature of the content itself. Is the level of the content in the resource suitable for the intended audience? If your library serves a user community composed primarily of undergraduate students, a graduate-level research database may not be the best resource to acquire. In evaluating the level of the material and its appropriateness for your constituency, consult with and take recommendations from faculty, staff, students, and whoever else is considered to be part of your primary audience.

One of the last issues to consider in terms of content criteria is where the resource is indexed. While the issue of indexing is important regardless of format, the question of where a resource is indexed is perhaps more important for an e-journal, where the journal is not physically present on the shelf and is unlikely to be browsed. If you are considering a subscription to any journal, it is always important to consider where users will find citations to the articles in the journal. If they do not find references to the journal in any of the citation databases to which your library subscribes, they are unlikely to discover and use this journal.

Design and Usability

Often the decision to purchase a resource comes down to a choice between content and usability—what the resource includes and how it functions. Design and usability selection criteria encompass the overall design of the interface, including factors such as functionality, ease of use, and administrative capabilities.

There are several areas of functionality to look for in a search interface. First and foremost the search operators the interface will accept should be considered. The ability to use Boolean operators is highly preferred. Users quickly become familiar with the Boolean operators "and," "or," and "not," and search interfaces should, if possible, accommodate their use. The ability to use other operators, such as "near," "with," and "within," is a bonus but probably not absolutely necessary.

The robustness of other search features carries weight in the decision whether to acquire an e-resource. Are both basic (simple) and advanced search options available? Can searches be combined, saved, and/or edited? The necessity of these more advanced search features is dependent upon the way users will be using the resource. If users are doing in-depth research on advanced concepts, looking to combine topics and to refine their searches, they will require more search features. If they are using the resource as a reference source, looking up a topic to verify a fact or find a single piece of information, a more robust search mechanism may not be necessary. When evaluating search features, it is useful to identify and seek out the search operators and features to which users at your institution are most accustomed.

Once a search has been performed and results are displayed, a new set of questions arises. Can the search results be refined, limited, or expanded? Can results be sorted, and, if so, how (i.e., by relevance, date, or other factors)? Again, the complexity of user needs will serve to define your criteria.

Once users have their data selected, whether it is a list of citations, a full-text article, or a chart of data, they will want to export and/or use that data in some way. The term "export" is used here in a broad sense, defined as any way the data can be extracted from the resource, whether it is by printing, saving, or copying and pasting content directly into a file. In evaluating the exporting capabilities of an e-resource, issues that you must investigate include the number of ways data can be exported and whether the resource can interface directly with bibliographic management software such as RefWorks and ProCite.

In addition to the functionality of the resource, it is important to consider the usability of the resource. How intuitive is the design? Can users easily figure out where to click and how to search, refine results, and export data? How much explanation is needed for users to be able to use the resource on their own? It is to be expected that some instruction on how to use a resource might be necessary. When evaluating usability, it is important to bear in mind that the steeper the learning curve, the less likely it is that users will want to utilize the resource. Often the more functionality a resource has, the more complicated it appears to the user. Oftentimes the challenge libraries face is striking the right balance between functionality and ease of use.

The last few criteria impacting the evaluation of the design and usability of a database are more for the librarian, as an administrator, than for end users. First, are usage statistics available, and, if so, are they COUNTER compliant? As users take advantage of online offerings and librarians are unable to observe firsthand exactly what it is that they are using (or not using), usage statistics are becoming increasingly important indicators of whether or not libraries are getting their money's worth out of a resource. COUNTER (http://www .projectcounter.org) compliance of usage statistics ensures that the same measures are being used across platforms and vendors, providing more reliable data.

In addition to usage statistics, some e-resource vendors provide access to an administrative module where you can set preferences for your institution. The preferences that you are able to set and define will vary from product to product and vendor to vendor. Common preferences that can be customized by an administrator include links to Interlibrary Loan (ILL) services, links to full-text resources and/or the online catalog, the number of results that are displayed at a time, and some search interface factors, such as whether or not the basic or advanced search is the default. In the absence of an administrative module, many of these preferences can be set or changed by contacting the vendor directly. In actuality, whether or not an administrative module is available and the preferences that can be customized within the module will probably play a relatively small role in the selection of a particular e-resource.

Accessibility

For the purposes of this chapter, the accessibility of a resource has two distinct components. The first component of accessibility is defined as how accessible the resource is to users on a technical level. The second component is defined as how accessible the e-resource is through interlinking with other resources and content.

When evaluating an e-resource, it is important to examine the technical requirements of the resource to identify any browser, software, server, and authentication requirements that may limit the library's ability to acquire and to provide access to a resource. "For cost considerations, the selector may have to ensure that the format of an electronic resource is compatible with the existing library hardware and software, unless there are funds available for simultaneous purchases of hardware and software."[1]

For example, does the resource require a certain browser level? Does it work well on multiple browser platforms? If the resource is not completely online, but instead has to be loaded on a local server, it is important to look at the server requirements as well. In addition to the basic hardware and software requirements, you will also want to determine if there are any proprietary software requirements. Does the resource require a proprietary viewer or reader to access the content? Some resources, such as e-book services and image archives, may require a user to download a specific program to be able to use the resource.

For online resources there are two primary authentication methods, IP address and username/password. "The username and password may be generic for the institution or linked to one or more specific individuals—sometimes one username and password will be shared by several people (e.g., an entire department)."[2]

IP address authentication matches the IP address of the user's computer with the permitted IP addresses for the user's institution. If the user is accessing the resource from an allowed IP address, that person is directly authenticated to access the resource. For those users attempting to access a resource from a remote location, an additional step of authentication such as a proxy server may be required. Some resources do not yet permit IP address authentication and may require the use of a username and password to access the resource.

Accessibility is also impacted by the linking into and out of a resource. The implementation of an OpenURL link resolver can help an end user find the full text of an article and identify other pertinent resources. However, users cannot take advantage of this technology if the e-resource is not OpenURL enabled. Accordingly, you should determine if the resource can send and receive links to and from a link resolver. In addition to linking to a link resolver, it is helpful to be able to set up direct links to other services. Some examples of other services to which you may want to provide links include full-text content in other resources such as JSTOR, an ILL module, or the online catalog. Be sure to investigate exactly how many links a resource allows you to display to a user and to what degree you can customize those links. In addition to linking out of the resource, you may find that you also want to provide direct links into the resource. For example, if the resource you are considering is a database of full-text journals, you may want to provide a title-level link to the journal content in your online catalog. Determine if the resource you are considering uses persistent URLs or a consistent linking syntax that will allow you and your patrons essentially to bookmark links to content.

Licensing

Reading the license agreement can be one of the most challenging and intimidating aspects of selecting and acquiring an e-resource. Determining if the terms of a license agreement meet your selection criteria cannot be done merely by checking items off on a list, and the process is complicated by the fact that some licensing terms are up for negotiation. For example, licensing terms such as ILL and online reserves, if not expressly permitted in the license agreement, will sometimes be allowed by the database provider upon request. Further complicating the negotiation is the fact that some criteria that may be up for negotiation such as the number of sites and simultaneous users that are permitted may also affect the price.

It is also important to note that several of the factors outlined in this licensing section also impact other selection criteria areas. For example, several of the terms in the license will affect the accessibility of the resource. How many sites does the license permit? If you have multiple campus locations or are negotiating a license for a consortium, this will be an important factor to consider. Equally important will be the number of simultaneous users who are allowed to access the resource. Frequently adding sites and users, which will increase accessibility, will typically also increase the price. A balance will need to be achieved between how much access to provide and the cost of the resource.

Another licensing term that will impact the accessibility of the resource is whether or not remote or off-campus access is permitted. Determine if there is a restriction on providing access to users who are not physically on the premises of your organization. If remote access is allowed, can that access be set up through a proxy server or is a username and password required?

Some licensing terms will affect the ability to share the content of e-resource subscriptions. Is ILL of full-text electronic materials permitted? If ILL is permitted, how can the content be transmitted? Are you allowed to transmit the content electronically, or are you required to print out the content and send it as a hard copy or facsimile? Similar questions can be asked regarding placing materials on reserve. Can the content be placed on reserve? If so, print or electronic? While these factors may not impact whether or not you choose to acquire a resource, they will certainly impact your ability to use the resource, and they are best addressed from the outset.

It is also important to determine whether the license agreement addresses issues of perpetual access. "The agreement should safeguard long-term access to all the material subscribed to, either by permitting copying for archival purposes, or, should the subscription cease, by allowing the institution to receive copies of all the material made available during the period of the license covered."[3]

In the event of cancellation, do you retain access to the years or content for which you have paid? If you retain access, how is that access provided? In some cases, vendors agree to provide access to the purchased content, but they provide it in the form of a compact disc that you are responsible for loading on a server and making searchable.

In addition to providing a guarantee of perpetual access, the vendor should make a commitment to archiving content. A recent report published by the Council on Library and Information Resources recommends that, "publishers should be overt about their digital archiving efforts and enter into archiving relationships with one or more e-journal archiving programs."[4] The report specifically highlights twelve different archiving initiatives, including LOCKSS and Portico. Has the vendor taken steps to archive its own content through a reputable initiative such as LOCKSS or Portico? If the vendor suffers a catastrophe that prevents it from being able to provide access, you will need some assurance that access will not be lost altogether.

It is important to remember that a license agreement works both ways; it is a mutual agreement between you and the vendor from which you are acquiring the e-resource. In addition to the terms outlined in previous text that you must either agree to or negotiate, there are elements of the license outlining the vendor's responsibilities to you as the customer and to the product it is selling. Determine if the vendor offers technical support and training and whether this support and training is free or comes at an additional charge. To what degree is the vendor available if you have a problem with the resource? In addition to offering technical support, the vendor is responsible for maintaining the resource and updating the files in a timely fashion. If a mistake is found in the resource or there is problem with a corrupt file, such as an out-of-date PDF, does the vendor make an assurance to correct these problems and in what time frame?

Many license agreements include some kind of confidentiality clause, requiring organizations to keep license terms or pricing, or both, confidential within the organization. These kinds of clauses restrict the librarian's ability to share pricing with colleagues at other institutions and can tie the institution's hands in terms of negotiation. Public institutions in "open-record" states are prohibited from signing license agreements with confidentiality clauses and are required to state their expenditure of state funds publicly.

This section has outlined some criteria that should be kept in mind while reading the license. You will have to determine which are the most important factors for your organization, and which points, if not met in the license, you are willing to negotiate or compromise on. By no means is this list of criteria meant to be comprehensive. There are a number of resources that can assist in your evaluation of a license agreement. The LibLicense Web site (http://www.library.yale.edu/~llicense/index.shtml) hosted by Yale University includes information on license terms and vocabulary as well as the archives of the LibLicense discussion list. The Association of Research Libraries provides access to licensing information and resources on its Scholarly Communications Web page (http://www.arl.org/osc/licensing/index .html). Licensingmodels.com (http://www.licensingmodels.com/), created and maintained by John Cox Associates, provides standard licenses for publishers, libraries, and subscription agents. For further information on licensing, see Chapter 15 by Jill E. Grogg.

Pricing

E-resources are typically priced in one of two ways, either as an annual subscription or as a one-time purchase. Materials that may be available as a one-time purchase include journal and newspaper backfiles and some e-books. The purchase price or one-time fee is typically more money up front but with a lower annual maintenance fee, if any at all. There are a few e-resources that do not require any additional expenditure of funds once a customer has paid the purchase price. Annual subscriptions typically have lower up-front costs than one-time purchases, but they are subject to market price increases and inflation.

If your library has multiple purchasing options, the kind of funds available will impact the choice to subscribe to a resource or to purchase it outright. If you have one-time funds or end-of-year money available, a one-time purchase using these funds, while more expensive upfront, may minimize or eliminate an annual expenditure of money. Annual maintenance fees for the purchase of an e-resource are typically much less than the cost of an annual subscription. If your library does not have the large chunk of money needed to purchase a resource or if the library is unwilling to make a long-term commitment to owning the resource, an annual subscription may be the more appropriate option.

A third pricing option is pay-per-use, where the fees are based upon the number of times the resource is used or searched. In this acquisition model, the more you search and/or access the resource's content, the more you pay. Pay-per-use can be a flat price per search, such as the way that Online Computer : Library Center (OCLC) bills FirstSearch searches, or it can be billed based upon the complexity of the search, such as the way that DIALOG bills for searches. This particular option is only effective if you are confident that the total of your per-search fees will be less than the cost of an annual subscription. This is most effective for resources that will be used infrequently or only for very specific subjects and circumstances.

If you are considering acquiring the resource through a consortium, investigate how the consortial pricing differs from the pricing for your library alone. Some vendors offer very favorable consortial arrangements, while others do not offer much of a consortial discount or do not work with consortia at all. Ordering a resource through a consortium adds an extra administrative level, and often license agreements are negotiated for an entire consortium. When acquiring a resource through a consortial partnership, you may need to sacrifice some licensing terms to get more favorable pricing.

Other pricing factors to take into consideration are the billing procedures and the inflation rate. Will the vendor allow third-party billing? Can you pay for a resource through your subscription agent? How many agents is the vendor willing to work with per institution? If ordering through a consortium, will the vendor bill through the consortium? In terms of the inflation rate, you want as much of a guarantee as possible from the vendor that the pricing will not change dramatically from year to year. Will the vendor offer a price cap on the inflation rate? If the vendor will place a price cap, are you required to make a commitment for more than one year to get it?

As previously indicated, depending upon licensing terms, your price may be partly determined by the number of users and sites for which the resource is licensed. As such, pricing, like licensing terms, is a factor that can be negotiated. To what degree the pricing is up for negotiation will depend upon the vendor. Some vendors absolutely refuse to negotiate on price, while others are fairly flexible.

CONCLUSION

All of the criteria outlined in the previous text have been incorporated into an E-resources Selection Criteria Worksheet, which can be found in Appendix 2.A. at the end of this chapter. Finding answers to all of the questions in the worksheet can be challenging. Listed in the annotated bibliography accompanying this chapter are a few resources that investigate e-resources and can help in the selection process. Also listed in the annotated bibliography are a few guides and manuals that offer advice on selection criteria and collection development policies for e-resources.

One of the most useful ways to determine if an e-resource is going to meet your users' needs and your selection criteria is to conduct a trial. Most vendors offer at least a thirty-day trial, and many are willing to extend the trial period. Trials will not usually be a success without publicizing them and directly soliciting feedback from interested parties.

It is important to remember that there are no hard-and-fast rules in the selection of e-resources. You will want to build a certain amount of flexibility into the selection process. It would be virtually impossible for an e-resource to meet all of the criteria outlined here. The key to successfully selecting the most appropriate resources for an organization is identifying the criteria that are most important for your organization, prioritizing these criteria, and then applying them to the e-resource(s) being evaluated. For example, at some organizations, it may not matter if a resource has a limited number of simultaneous users, while at others it may be imperative to have unlimited users. The list of prioritized criteria can be used to form a comprehensive selection policy, encompassing both print and e-resources.

NOTES

1. Vicki L. Gregory, *Selecting and Managing Electronic Resources,* with assistance by Ardis Hanson, "How-To-Do-It Manuals for Librarians" 146 (New York: Neal-Schuman Publishers, Inc., 2006), 19.

2. Stuart D. Lee, *Electronic Collection Development: A Practical Guide* (New York: Neal-Schuman Publishers, Inc., London: Library Association Publishing, 2002), 23.

3. Ibid., 92.

4. Anne R. Kenney, Richard Entlich, Peter B. Hirtle, Nancy Y. McGovern, and Ellie L. Buckley, *E-Journal Archiving Metes and Bounds: A Survey of the Landscape,* CLIR Publication 138 (Washington, DC: Council on Library and Information Resources, September 2006), 2, http://www.clir.org/PUBS/reports/pub138/pub138.pdf (accessed December 19, 2006).

APPENDIX. ELECTRONIC RESOURCE
SELECTION CRITERIA WORKSHEET

1. Preliminary Questions
   a. Is the e-resource intended to be a replacement for or a supplement to a print subscription?
   b. What is the primary need this resource is supposed to fill?
   c. How will users use the resource?
2. Content
   a. What title(s)/item(s) are indexed in the resource?
   b. Are abstracts provided?
   c. What content is available in full text?
   d. How much overlap duplication is there with your existing subscriptions?
      1) E-resources?
      2) Print?
      3) Microform?
   e. If full text is available, is there cover-to-cover access to the contents?
      1) Are figures and tables included?
      2) Are graphics included?
   f. How is the content made available?
      1) PDF page images?
      2) HTML?
      3) Other?
   g. Is there an embargo on any content?
   h. How quickly is new content added after publication?
   i. How frequently is the resource updated with new content?
   j. Is the subject material and level of the content appropriate for the audience?
   k. Where is the resource itself indexed? Where will your users find references to the resource?
3. Design and Usability
   a. How functional is the search interface?
      1) Does it use Boolean operators?
      2) Can searches be combined?
      3) Can searches be saved and/or edited?
      4) Are both basic and advanced searches available?
   b. Can searches be refined / limited / or expanded?
   c. Can results be sorted?
      1) Date?
      2) Relevance?
      3) Author?
      4) Journal/Source?
      5) Other?

      d. What forms of data export/output are there for search results?
         1) Printing?
         2) Saving?
         3) Copy and paste?
4. Direct export to bibliographic management software?
      a. How intuitive/easy to use is the interface? How much explanation is required to use the resource?
      b. Are usage statistics available? Are the usage statistics COUNTER compliant?
      c. Is there an administrative module where you can set preferences for your institution?
5. Accessibility
      a. Technical requirements
         1) Is a proprietary reader or piece of software required?
         2) Are there any specific browser requirements?
            a) Browser level?
            b) Support multiple browser platforms (i.e., Internet Explorer, Mozilla, etc.)?
         3) Are there any server requirements?
         4) How is access to the resource authenticated?
            a) IP address authentication?
            b) Username and password?
      b. Linking
         1) OpenURL enabled?
         2) Direct linking to alternative resources (i.e., full-text content, ILL module, online catalog)?
            a) How many links can be set up?
            b) Can the links be customized?
         3) Persistent URLs linking to content?
6. Licensing Issues
      a. How many sites does the license permit?
      b. How many simultaneous users are allowed at one time? Unlimited?
      c. Can access be provided remotely to users off campus?
         1) Proxy Server?
         2) Username and password?
      d. Is interlibrary loan of full-text electronic materials permitted?
         1) Can the content be transmitted electronically via e-mail or a system like Arial?
         2) Does the content have to be transmitted in hard copy, such as fax or by mail?
      e. Can electronic materials be placed on reserve?
         1) In print?
         2) Electronically?

    f. If the subscription is cancelled, does the library retain access to the years for which it paid? How?

    g. In addition to assurances of perpetual access, has the vendor made a commitment to archive its content via a reputable initiative such as a Portico or LOCKSS?

    h. Does the vendor offer free technical support and training? Does the vendor agree to maintain and repair problems with files (e.g., with PDFs) in a timely manner?

    i. Does the vendor require that certain licensing terms be kept confidential?

7. Pricing
    a. Cost of an annual subscription?
    b. Purchase price of the resource?
    c. Annual maintenance fee?
    d. Pay-per-use?
    e. Does the vendor offer consortial pricing?
    f. Does the vendor allow you to pay for a resource through a third-party?
    g. Does the vendor offer a price cap on the inflation rate? Does the price cap require that the library enter into the agreement for more than one year?

Chapter 3

Preservation Concerns
in the E-Resource Environment

Jennifer Watson

INTRODUCTION

Preservation means many things to many people. Ask any group of librarians to name the issues surrounding preservation today and you will get many different answers. One might talk about the need to preserve print and archival materials through digitization, microfilming, or binding. Another might be concerned with the preservation of digital materials, the fear of changing technology, data degradation, and the vagaries of the commercial publishing world. A third might welcome the use of institutional repositories as a means to preserve, organize, and publicize material produced by an institution. Someone among the group is bound to raise the issue of continued access to materials when subscriptions are cancelled.

The preservation of electronic resources is a hot topic. It is discussed in articles, white papers, conferences, and mailing lists. The June 2006 issue of *Serials Review* was almost entirely devoted to this subject. With preservation encompassing such a broad range of issues, it can be hard to step back far enough to see the whole picture. With the technological and publishing landscape changing so quickly, it is equally difficult to predict the future. This chapter will explore the major issues affecting the preservation of electronic resources, especially electronic journals, and examine ways in which libraries, publishers, and others are trying to ensure that these resources are available in both the near- and long-term future.

THE CHALLENGES OF E-RESOURCE PRESERVATION

Traditionally, libraries collected materials in predominately print formats. By relying upon binding, book repair, climate control, anti-theft devices, legal deposit systems, and microfilming, libraries have developed an imperfect but largely effective infrastructure for preserving their print collections. As libraries make the transition from print to electronic materials, they enter a new preservation landscape. While some aspects of print preservation techniques can be applied to e-resources, in other ways the management of these resources presents new and unique challenges.

Maintaining Perpetual Access

A central concern for a library making the transition from print to e-resources is maintaining perpetual access to subscribed content. "When the subscription to the print version of a journal is cancelled, the library retains all the earlier issues of the journal to which it subscribed. With the online version, this may not be the case. A library cancelling an online subscription may lose all access to the journal, including issues for which access had previously been paid."[1] While this problem is primarily discussed in terms of journals, it can also affect any electronic content, much of which is accessed through a subscription rather than a one-time fee. Raising the stakes of the perpetual access problem is the stipulation in many license agreements limiting or even prohibiting the interlibrary loan of electronic content; the result is that it becomes difficult for a library losing access to obtain material from elsewhere.[2]

Despite the dangers of losing perpetual access, in practice it does not appear that perpetual access concerns have had any significant impact on libraries' decision making in terms of the models and licenses they rely upon to provide their users with access to e-resources. For example, in a recent informal survey of librarians' response to a hypothetical 50 percent budget cut, cancellation of expensive, underused, or duplicate subscriptions ranked high in the responses, but no one mentioned a lack of perpetual access rights as a concern when considering cancellation.[3] Indeed, some publishers "decline license requests for perpetual access on the basis that the demand from other institutions does not exist."[4] Moreover, some libraries are moving away from subscriptions altogether. Indeed, while librarians often use the term "access versus ownership" to describe the differences between online and print subscriptions, the increased use of pay-per-view and document delivery services for less-used titles is shifting the way in which even access is viewed. Libraries using these services would not even think about perpetual access—these are individual articles that are ordered at the time of need. While traditionally accreditation bodies and others have viewed some kind of ownership of content, even just leased, as a standard for libraries, this is starting to shift. For example, hospitals used to be required by the Joint Commission on Accreditation of Healthcare Organizations (the United States' main accreditation body for health care facilities) to maintain a library. Now, only access to information is required.[5] Just-in-time collection development is on the rise, and just-in-time delivery of content is becoming increasingly attractive to libraries.

So, why are libraries willing to compromise their desire for perpetual access? There are several answers to this question.

Patron Pressure. One force driving libraries to ignore the dangers of not having perpetual access is pressure from patrons. Patrons want access at their desktops, in their office, and at home. They are reluctant to interrupt their workflow by stopping what they are doing to visit the library in the hope that the needed article is on the shelf. With the rise in distance education, visiting the library is not even an option for many patrons. While patrons may be aware of the importance of archiving in the abstract,[6] their day-to-day concern is with current access to information. Librarians unable to meet these expectations may find themselves at odds with their patrons. This leaves them disinclined to further antagonize the relationship by turning down a request because the vendor does not offer perpetual access.

The Availability of Content. For the time being, a library usually has the option to subscribe to content in print. However, in future years, enhanced content may make the print version less and less of a replica of the content of note. In addition, born digital content will probably make up a larger percentage of the material libraries collect. For example, the British Library 2005 predicts a switch from print to digital publishing by the year 2020, with 90 percent of British research monographs being published in electronic format by that date.

Financial Pressures. Financial pressures come in a number of forms. It is a widely known fact that libraries' budgets have not increased in line with the rising cost of their subscriptions.[7] Meanwhile, an increasing number of journals and books are available in online as well as print format. New electronic products, such as databases, are constantly coming on the market. As libraries look to stretch their budgets to pay for these new resources, many see cancelling print subscriptions and instead relying upon online access as a way to save money.[8]

In addition to decreasing subscription costs, relying upon online-only access eliminates the cost of storing a physical item. A lack of shelving space is a major concern for libraries. Many libraries just do not have enough room to continue expanding their print holdings the way they could in past years. Expanding the library building is cost prohibitive in the current budgetary climate. Off-site storage is a more affordable option, but, as patrons become accustomed to immediate online access, they are less inclined to wait hours or days for requested materials to arrive from elsewhere. Just maintaining the current print collection is expensive in terms of binding, staffing, and maintenance.[9]

Finally, these financial limitations often translate into human resource limitations. While publishers may be amenable to adding wording to licenses related to perpetual access, license negotiations take time and libraries may not have the staff time available to negotiate perpetual access clauses into every contract. Besides, the typical vaguely worded commitment to perpetual access that can be negotiated into contracts, along with lack of evidence in how this wording translates into action, may make librarians wonder if the time spent negotiating such clauses into contracts is worth the effort. In addition, such clauses often become worthless when the content is purchased by another publisher. Ed Shreves of the University of Iowa Libraries described these clauses as "convenient fictions."[10] If perpetual access takes the form of local hosting or mounting, in the form of tapes or CDs, for example, the library may not have the staff available to set up and run this access.

Librarians must work to provide the best perpetual access options they can, while staying within limited budgets. While some vendors offer better perpetual access rights than others, libraries may select content that is less desirable from a perpetual access standpoint in order to meet patron needs or stay within budget. For instance, aggregator content is often more affordable, but few aggregators offer perpetual access rights.[11] Libraries can rarely afford to pay extra or purchase duplicate content purely for perpetual access reasons. In addition, some vendors, such as Ovid, require that the library continue to pay an access fee for the platform.[12] Nature Publishing Group revised its policy to provide for continued access to cancelled titles after the California Digital Library refused to sign an agreement without guarantees of perpetual access, but, like Ovid in previous text, a fee is required for this access.[13] Libraries cancelling titles due to a budget reduction may have difficulty finding funds for continued access to cancelled or discontinued titles. Ultimately, there is a delicate balancing act between perpetual access on the one hand, and patron demands and cost considerations on the other.

Data Copying and Migration

In a print dominated environment, a library could often preserve a unique or at risk item in its collection by photocopying it or creating a back-up copy of the item on microfilm. Today, however, libraries are increasingly collecting digital materials that cannot be easily reproduced in print or on microfilm. These materials may include video, audio, or three-dimensional content, as well as large datasets accompanying published research that would be prohibitive to include in a print copy.[14] Moreover, hyperlinks to related materials (that cannot be replicated in print materials) are becoming an integral part of e-journal articles.[15]

As a result, libraries are preserving e-resource content using digital mediums. Carrying out this task is a major challenge. One reason for this is the cost involved. For example, a recent study published by the Ricksarkivet (National Archives) in Sweden found that "the costs of digital storage are much higher than generally believed."[16] Funding must be found not just for the hardware and software needed, but also to hire staff with the technical expertise necessary to create and maintain archiving systems. A second reason for the challenge of preserving content on digital mediums is the maintenance and migration of content. Data stored ten years ago may have been created using software that is no longer available or that has changed so dramatically that backward compatibility is impossible. Degradation of the storage medium is also a concern. Tapes and compact discs may become damaged over time or simply wear out. Data on hard drives can become corrupted.[17] Storage devices change over time; for instance, modern machines are not able to read 1.2 inch floppy disks.[18] Migrating from one format to another is labor-intensive, time-consuming, expensive, error prone, and risky.[19] On a more encouraging note, however, the storage capacity of computers is increasing rapidly. An unwieldy dataset of today will probably be easy to manage in a few years.[20]

Preserving Content in the "Information Explosion"

With the transition from print to e-resources, the world has experienced what many have termed an "information explosion": a profound increase in the creation, modification, and distribution of digital content. A large portion of the content contributing to the information explosion takes the form of communication in electronic arenas. Traditionally, communications in print formats have often been archived for study, as well as for administrative and legal compliance purposes. With electronic communications, however, this is not always the case. Indeed, although Web sites, e-mails, blogs, Wikis, chat, podcasts, and instant messaging (IM) are all important sources of information that studies indicate the value of retaining,[21] libraries have only given limited consideration to their preservation.

This is unfortunate given how difficult the preservation of these communications can be. For example, a nonprofit group, Internet Archive, has been archiving the Internet since 1996. While its current archive, known as the Wayback Machine, does provide some archiving of Web sites, it does so only periodically, and only tracks Web sites that are readily accessible by automated crawlers.[22] Content that changes frequently, or is inaccessible to crawlers due to password protection, etc., will not be tracked effectively by the Wayback Machine, if at all. Blogs and Wikis by their nature change frequently and so are not good candidates for being archived in this manner. Blogs are extremely difficult to archive. There is a huge volume of content that changes frequently. In addition, the content is often highly intertextual, ephemeral, and personal, so an archived version of a blog may be seen as being of limited use.[23]

Another form of digital content that is just as dynamic as blogs and Wikis and has also received little attention concerning preservation is reference works. As these works go online, they have evolved from static print materials to digital content that is constantly updated. As a result, libraries formerly subscribing to and retaining print reference materials are losing the ability to use them for research. As Hartman and D'Aniello comment, "superseded reference works . . . [capture] a snapshot at a point in time."[24] Superseded directories are used by researchers studying history, as well as business and medical trends. Moreover, when researchers cite an online reference work, once the work is updated it becomes impossible for future researchers to verify the citation.

Even if preservation techniques are developed that accommodate the increasingly dynamic nature of digital content, libraries will not be able to preserve all of it. Instead, libraries must be selective; as Lavoie and Dempsey write, "preserving everything is not an option."[25] The Digital Preservation Coalition has published a selection tool to aid in the decision-making process. The tool assists the user in determining whether the resource has sufficient long-term value to justify preservation, as well as taking into account the availability of the resource in other formats, the preservation responsibilities of the organization, and technical considerations.[26]

PARTNERSHIPS AND INITIATIVES
IN E-RESOURCE PRESERVATION

The previous section focused on the challenges of e-resource preservation, the technical considerations, the dearth of initiatives relating to the preservation of new forms of communication, and the problems libraries face in ensuring continued access to cancelled materials. This section provides an overview of how the various and interrelated stakeholders are working to ensure that selected electronic resources are preserved for future generations. Third-party organizations are working with publishers and libraries to preserve content, particularly electronic journal content, on their behalf. Government legislation is being updated to reflect the widespread use of electronic means of information dissemination, and foundations are funding projects to investigate and find solutions for electronic preservation issues.

Third Parties

Third-party organizations are currently providing some of the most promising models for preservation. Chief among these organizations are JSTOR, LOCKSS, and Portico.

JSTOR. The first of these organizations, JSTOR, is a subscription-based service that provides access to back issues of scholarly journals. Libraries pay a fee to subscribe to the service that can be purchased on a collection-by-collection basis. Purchase of individual titles is not permitted and no perpetual access rights are granted in the event that a subscription is cancelled.[27] JSTOR aims to supplement but not replace subscriptions to current content, and, to this end, there is a rolling embargo, the length of that varies from publisher to publisher. The length of the embargo is significant: Blackwell has announced that its JSTOR embargo will be increasing from three to ten years.[28] Presumably, Blackwell's reasoning is that the shorter embargo was hurting sales of its current content. While many libraries are wary of cancelling print subscriptions, JSTOR seems to have built a level of trust with libraries beyond that of other online vendors. A recent survey of seventeen subscribing libraries found that eleven of them had discarded print volumes of content available through JSTOR. JSTOR is further enhancing its image as a reliable provider of content by working with the University of California and Harvard University to create print archives of all of its online holdings.[29]

LOCKSS. Another important third-party organization, Lots Of Copies Keep Stuff Safe (LOCKSS), aims to mirror the library's traditional preservation role by distributing electronic preservation efforts across all of the participating libraries. Each library maintains a LOCKSS server that stores content from materials that are freely available or subscribed from participating publishers. Before archiving content, each library must negotiate this right into its license with each participating publisher.[30] The servers can provide backup to each other in the event that the data on one library server becomes corrupted, and can provide content to participants if it becomes unavailable online. LOCKSS differs from other archiving initiatives in a number of ways:

1. Its distributed environment, as outlined in previous text, reduces the risk of loss due to media failure, hacking, or tampering.[31]
2. The instant accessibility of stored content, without reference to limited "trigger events."
3. Its ability to be used as a means of providing perpetual access to libraries cancelling subscriptions.
4. When accessed, content is indistinguishable from that on the publisher's site, and shows the content in its original (online) context.[32]

However, presenting the content in the publishers' format can also be seen as a deficiency, as there is a danger of the data format becoming obsolete over time.[33] As content from LOCKSS is served during any outage of the publisher's Web site, LOCKSS has had plenty of opportunities since its inception in 1999 to prove that it works. No other e-journal archiving system has had as much real-world testing.[34]

There are currently seventy publishers participating and 140 LOCKSS "boxes."[35] LOCKSS' low cost makes it an attractive option for smaller publishers.[36] A new version of LOCKSS is Closed LOCKSS (CLOCKSS), a dark archive in which the content becomes available only if an online journal has been inaccessible for at least six months. Like LOCKSS, CLOCKSS relies on a distributed model of administration.[37]

Portico. Portico shares its heritage with LOCKSS in that both were started with funding from the Mellon Foundation. However, Portico takes a different approach from LOCKSS, working from the assumption that libraries do not possess, or want to acquire, the infrastructure, funding, and personnel necessary for long-term storage of digital resources. Originally an initiative of JSTOR, Portico gained its name and independence in 2005.[38] Portico's centralized archive is "dark," or unavailable, until a publisher stops operations, ceases to publish a title, stops offering back issues, or "upon catastrophic and sustained failure of a publisher's delivery platform."[39] Portico normalizes data to ensure that it can be easily migrated as file format standards change and plans to audit content to ensure it does not become corrupted over time. Like JSTOR, Portico does not accept post-publication corrections.[40] Participating publishers pay annual fees ranging from $250 to $75,000.[41]

As of April 2006, nine publishers had made commitments to contribute a total of 3,200 journals to the archive.[42] However, in a presentation to the University of Tennessee Libraries on September 28, 2006, Portico's Bruce Heterick admitted that little content had been loaded into the archives as of yet. Library fees, ranging from $1,500 to $24,000 per annum, were introduced in 2006. While Fenton claims that "nearly three dozen institutions [are] committed to supporting the Portico archive,"[43] no library participants are named on the Portico Web site. This is in stark contrast to the listing of publishers and journal titles, and suggests that Portico may not, as of yet, count many libraries as active, paid-up participants. However, Portico may present an attractive option to libraries not wishing to maintain their own LOCKSS servers or for publishers concerned about lack of control over content on LOCKSS servers.[44]

Other Third Parties. In addition to JSTOR, LOCKSS, and Portico, there are a number of other third-party preservation initiatives that are of note. One example is Google Book Search (originally called Google Print), which was launched in December 2004. The project involves digitizing books from Michigan, Stanford, Harvard, and Oxford universities, as well as the New York Public Library. Works out of copyright are freely available and searchable online.[45] In addition, Google is now inviting publishers to submit materials for digitization.[46] University of Michigan President, Mary Sue Coleman, has described the project as "one of the most extensive preservation projects in world history." The project complements ongoing work at the university to digitize brittle and damaged materials, and the digitized texts provide a facsimile in the event that the original printed material is destroyed.[47]

Another notable third-party initiative, PubMed Central, provides selected online life sciences journal content free of charge and would form a backup copy in the event of failure at the publisher site. The archive, managed by the United States National Institutes of Health's National Center for Biotechnology Information, is expanding as earlier print issues are scanned and added to the Web site.[48]

Libraries

These initiatives have not been limited to third parties. Preservation is also a key component of libraries' missions,[49] but many are unsure as to how this can be achieved.[50] As mentioned in previous text, some libraries are participating in LOCKSS. Other libraries are retaining print versions of online journals, or participating in cooperative print retention initiatives. For instance, member libraries of the Committee on Institutional Cooperation's (CIC) Center for Library Initiatives have banded together to purchase a single copy of journal titles published by Springer Verlag KG and John Wiley & Sons. These journals are stored in two CIC off-site shelving facilities, where environmental controls are designed for optimal preservation of the materials.[51]

In addition, many libraries are establishing institutional repositories where material produced by researchers within the institution can be stored. However, a survey of 340 librarians revealed that many are unsure of the completeness and long-term integrity of repositories and feel that preprints and postprints cannot substitute for the published journal article.[52] Indeed, oftentimes the material published here is not identical to the final published version, which is a practice that can result in discrepancies as other researchers review these alternative versions of the article.[53] Morris, for example, speculated that there could be more than a dozen versions of an article as it moved through the publication process.[54] Lastly, it can be difficult to convince senior management within institutions of the value of, and need to invest in, digital preservation projects such as institutional repositories.[55] This is especially true given the difficulty of convincing authors to enter publications in institutional repositories; for example, a 2004 survey found that the average institutional repository contained only 1,250 documents.[56]

On the national level, libraries are also taking action to preserve e-resources. In 2002, a joint steering group of the International Federation of Library Associations and Institutions (IFLA) and the International Publishers' Association (IPA) recommended that national libraries should take responsibility for long-term digital preservation.[57] Koninklijke Bibliotheek (KB, or National Library of the Netherlands) is the most well-known example of a national library initiative. Its e-Depot has made it a leader in preservation initiatives. In 1999, the

Dutch Publishers Association agreed to deposit electronic publications with the KB, and similar agreements have been made with many leading publishers including Elsevier Science, BioMed Central, Blackwell Publishing, Oxford University Press, Taylor & Francis, SAGE Publications, and Springer Verlag KG.[58] Content from the e-Depot is available to registered users while within the KB, via interlibrary loan with the Netherlands, and to any licensee where the content becomes unavailable due to publisher bankruptcy or other catastrophic circumstances.[59]

Publishers

Under pressure from libraries, publishers are also actively pursuing preservation opportunities. A 2004 study found that 52 percent of commercial publishers and 45 percent of nonprofit publishers had made formal preservation plans.[60] However, questions remain about how reliable publishers are as a solution to long-term preservation needs. Publishers are bought, sold, and go out of business. Their main concern is profit rather than preservation, and they have no experience in this field.[61] Many publishers, in looking to scan older issues of journals to make available in digital form, have had to look elsewhere for copies, as they did not retain complete backfiles of the journals themselves. For instance, when Oxford University Press began to digitize its backfiles, the publisher only held a small proportion of the issues needed. The rest had to be purchased from back-stock agents, borrowed from learned societies, or sourced from the British Library.[62]

Recognizing libraries' concerns over long-term preservation and distrust of publishers' abilities to provide solutions, publishers are turning to third-party providers perceived as having the ability to maintain archives in the long term. LOCKSS and Portico are popular archival service providers: SAGE Publications, John Wiley & Sons, and Oxford University Press are among publishers signing up with one or both of these services. JSTOR lists more than 400 publishers on its Web site as having reached an agreement to provide content for its Web site. Many publishers are aligning with more than one archive provider. For example, Elsevier has reached an agreement with Koninklijke Bibliotheek[63] and Portico to archive its journal content. Figure 3.1 shows publishers who have, or plan to, provide content to more than one of

- African Studies Association (JSTOR and LOCKSS)
- American Anthropological Association (JSTOR and Portico)
- American Folklore Society (JSTOR and LOCKSS)
- American Historical Association (JSTOR and LOCKSS)
- American Mathematical Society (JSTOR and Portico)
- American Society of Plant Biologists (JSTOR, LOCKSS, and PubMed Central)
- Annual Reviews (JSTOR and Portico)
- Association of Schools of Public Health (JSTOR and PubMed Central)
- Berkeley Electronic Press (LOCKSS and Portico)
- BioMed Central (Koninklijke Bibliotheek and PubMed Central)
- BioOne (LOCKSS and Portico)
- Blackwell Publishing (JSTOR, Koninklijke Bibliotheek, and PubMed Central)
- Brill Academic Publishers (JSTOR and Koninklijke Bibliotheek)
- Brookings Institution Press (JSTOR and LOCKSS)
- Cell Stress Society International (JSTOR and PubMed Central)
- Duke University Press (JSTOR and LOCKSS)
- Edinburgh University Press (JSTOR and LOCKSS)
- Elsevier (Koninklijke Bibliotheek and Portico)
- George Washington University Institute for Ethnographic Research (JSTOR and LOCKSS)
- Hastings Center (JSTOR and LOCKSS)
- Institute of Physics (LOCKSS and Portico)
- Johns Hopkins University Press (JSTOR, LOCKSS, and Portico)
- John Wiley and Sons (JSTOR and Portico)
- MIT Press (JSTOR and LOCKSS)
- National Academy of Sciences (JSTOR and PubMed Central)
- National Institute of Environmental Health Science (JSTOR and PubMed Central)
- Ohio State University Press (JSTOR and LOCKSS)
- Oxford University Press (JSTOR, Koninklijke Bibliotheek, LOCKSS, Portico, and PubMed Central)
- Princeton University Press (JSTOR and LOCKSS)
- SAGE Publications (JSTOR, LOCKSS, and Portico)
- Society for Industrial and Applied Mathematics (JSTOR and Portico)
- Springer (JSTOR and Koninklijke Bibliotheek)
- Taylor & Francis (JSTOR and Koninklijke Bibliotheek)
- University of Chicago Press (JSTOR and Portico)

FIGURE 3.1. Publishers Providing Content to Multiple Archiving Services

the following services, according to data compiled from respective services' Web sites: JSTOR,[64] Koninklijke Bibliotheek,[65] LOCKSS,[66] Portico,[67] and PubMed Central.[68]

Governments

In many countries (including France, Greece, Indonesia, Norway, Peru, South Africa, Sweden, Australia, Great Britain, the United States,

Canada, Japan, Nigeria, and Venezuela),[69] the government requires a publisher to submit a print copy of all its publications to a legal depository. For instance, in the United Kingdom and Republic of Ireland, one copy of any printed material distributed within either country must be given to the British Library in London and, on request, to each of five deposit libraries.[70] In some countries where legal deposit of print materials is required, governments have sought to extend legislation to cover electronic materials. In the United Kingdom, the Legal Deposit Libraries Act of 2003 established a framework for extending existing legal deposit legislation to cover nonprint materials.[71] Moreover, in the United States, researchers receiving funding from the National Institutes of Health (NIH) are advised to deposit their articles at PubMed Central after one year. Compliance with this voluntary recommendation has been low, and the NIH is considering beefing up the deposit system, possibly by mandating deposit.[72]

Foundations

Nonprofit foundations have been active in recent years in funding research into preservation issues. For instance, in 1995 the Andrew W. Mellon Foundation founded and funded JSTOR.[73] Five years later, the same foundation funded seven studies that examined various aspects of digital preservation.[74] The Hewlett Foundation issued a grant to the Internet Archive in 2006 to fund its digital preservation work.[75] In the United Kingdom, the Wellcome Trust has funded a Web archives feasibility study (since 2002) and a project, begun in 2004, to digitize and make available a number of medical journals.[76]

CONCLUSION

As this chapter has demonstrated, the challenges of e-resource preservation are many, and solutions just in their infancy. There continues to be much debate over the issue, but a lack of funds, uncertainty over the future, and who should take responsibility for preservation have resulted in a patchwork of preservation attempts by publishers, libraries, and others. Third-party providers such as JSTOR, LOCKSS, and Portico are working to fill the void, particularly in terms of electronic journals, but it remains to be seen how well these types of initiatives will be able to attract funding and prevent data obsoletion and corruption over the long term. Other efforts that are having a positive impact on e-resource preservation are government legislation and grant funding from nonprofit organizations. However, much more needs to be done, especially with regard to nonjournal resources, which have received little attention to date.

Libraries are driven by a shortage of funding and space, and patron demand for desktop access, to cancel and discard print journals in favor of online, but this same lack of funding restricts their ability to invest in or support preservation initiatives. As a result, libraries are increasingly finding that they are renting journal content, rather than owning it. If libraries are to continue their traditional role as preservation leaders, then their parent institutions must recognize the need to support preservation initiatives financially. Another way in which institutions can support library preservation efforts is by putting policies in place that encourage or require researchers to deposit content in institutional repositories.

Electronic resources provide many benefits. Materials can be accessed instantly any time and from anywhere, which is ideal for distance learners and those with busy schedules. Unlimited online access means that the item is never checked out to another user, lost, or out for repair. Multimedia content and datasets enhance the presentation of research and aid visual learners. However, librarians should be wary of rushing headlong toward these new opportunities without weighing all the factors involved. The wealth of materials from the past to which researchers have access today is due in large part to the careful preservation efforts of librarians and others through the ages. If a cohesive preservation strategy for electronic resources is not devised and implemented in the near future, there is a real danger that current research output may be lost to future generations.

NOTES

1. Jennifer Watson, "You Get What You Pay For? Archival Access to Electronic Journals," *Serials Review* 33, no. 3 (2005): 200.

2. Jim Stemper and Susan Barribeau, "Perpetual Access to Electronic Journals: A Survey of One Academic Research Library's Licenses," *Library Resources & Technical Services* 50, no. 2 (2006): 104.

3. Rick Anderson, e-mail to SERIALST mailing list, March 10, 2006.

4. Stemper and Barribeau, "Perpetual Access," 103.

5. Andrew Paradise, "Why the Joint Commission on Accreditation of Healthcare Organizations Should Add New Regulations Regarding Libraries," *Journal of the Medical Library Association* 92, no. 2 (2004): 167. http://www.pubmedcentral.nih.gov/articlerender.fcgi?tool=pubmed&pubmedid=15088075 (accessed August 13, 2006).

6. Stemper and Barribeau, "Perpetual Access," 91-92.

7. Richard E. Quandt, "Scholarly Materials: Paper or Digital?" *Library Trends* 51, no. 3 (2003): 350.

8. Patsy Baudoin, "Uppity Bits: Coming to Terms With Archiving Dynamic Electronic Journals," *The Serials Librarian* 43, no. 4 (2003): 66-67.

9. C.H. Montgomery and J. L. Sparks, "The Transition to an Electronic Journal Collection: Managing the Organizational Changes," *Serials Review* 26, no. 3 (2000): 10.

10. Stemper and Barribeau, "Perpetual Access," 94.

11. Ibid., 103.

12. Ovid Technologies, Inc., "Journals@Ovid Electronic Archive Policy," http://www.ovid.com/site/products/journals_archive_policy.jsp?top=2&mid=3&bottom=7&subsection=12 (accessed April 29, 2006).

13. Stemper and Barribeau, "Perpetual Access," 94, 103.

14. Baudoin, "Uppity Bits," 66.

15. Peter B. Boyce, "Who Will Keep the Archives? Wrong Question," *Serials Review* 26, no. 3 (2000): 54-55.

16. Jonas Palm, "The Digital Black Hole," http://www.tape-online.net/docs/Palm_Black_Hole.pdf (accessed May 7, 2006).

17. Sally A. Buchanan and Kirsten Jensen, "The Electronic Link," *Wilson Library Bulletin* 69 (1995): 53-54.

18. Terry Kuny, "A Digital Dark Ages? Challenges in the Preservation of Electronic Information," http://www.ifla.org/IV/ifla63/63kuny1.pdf (accessed May 7, 2006).

19. Stewart Granger, "Emulation as a Digital Preservation Strategy," *D-Lib Magazine* 6, no. 10 (2006), http://www.dlib.org/dlib/october00/granger/10granger.html (accessed March 16, 2006).

20. Jim Gray, Alexander S. Szalay, Ani R. Thaker, Christopher Stoughton, and Jan vandenBerg, *Online Scientific Data Curation, Publication, and Archiving* (Redmond, WA: Microsoft Research, Microsoft Corporation, 2002), http://arxiv.org/pdf/cs.DL/020812 (accessed May 10, 2006).

21. Christine A. Halverson, "The Value of Persistence: A Study of the Creation, Ordering and Use of Conversation Archives by a Knowledge Worker," *Proceedings of the 37th Hawaii International Conference on System Sciences – 2004* (2004): 9. http://csdl2.computer.org/comp/proceedings/hicss/2004/2056/04/205640108a.pdf (accessed May 24, 2006).

22. Internet Archive, "Internet Archive Frequently Asked Questions," http://www.archive.org/about/faqs.php (accessed May 22, 2006).

23. Alex Halavais, "Blogs and Archiving," http://alex.halavais.net/news/index.php?p=825 (accessed May 23, 2006).

24. Donald K. Hartman and Charles A. D'Aniello, "Subscribe to an Online Directory Today, Frustrate a Researcher Tomorrow: Are Print Directories Dead?" *College & Research Libraries News* 67, no. 4 (2006): 222.

25. Brian Lavoie and Lorcan Dempsey, "Thirteen Ways of Looking at . . . Digital Preservation," *D-Lib Magazine* 10, no. 7/8 (2004), http://www.dlib.org/dlib/july04/lavoie/07lavoie.html (accessed March 16, 2006).

26. Digital Preservation Coalition, "Interactive Assessment: Selection of Digital Materials for Long-Term Retention," http://www.dpconline.org/docs/handbook/DecTree.pdf (accessed April 30, 2006).

27. Stemper and Barribeau, "Perpetual Access," 100.

28. Van Sickle, e-mail to SERIALST mailing list, April 17, 2006.

29. John Kiplinger and Rebecca Kemp, "Print Backfiles in the Age of JSTOR," Paper presented at the *Fifteenth North Carolina Serials Conference,* Chapel Hill, North Carolina, March 30-31, 2006.

30. Stemper and Barribeau, "Perpetual Access," 101.

31. M. Seadle, "A Social Model for Archiving Digital Serials: LOCKSS," *Serials Review* 32, no. 2 (2006): 74.

32. Ibid., 76.

33. Marilyn Geller, "Models for E-journal Archives: Future Pathways into the Past," In *The E-resources Management Handbook,* United Kingdom Serials Group, 50, http://uksg.metapress.com/link.asp?id=ef0xtn27hq6n5myu (accessed August 7, 2006).

34. Seadle, "A Social Model," 77.

35. Cris Ferguson, "ATG Interviews Victoria Reich," *Against the Grain* 18, no. 2 (2006): 50.

36. Stemper and Barribeau, "Perpetual Access," 94.

37. V. Reich, "Follow the Money!" *Serials Review* 32, no. 2 (2006): 68-69.

38. Geller, "Models for E-journal," 50.

39. E.G. Fenton, "An Overview of Portico: An Electronic Archiving Service," *Serials Review* 32, no. 2 (2006): 83.

40. Geller, "Models for E-journal," 51.

41. Fenton, "An Overview of Portico," 84.

42. Eileen Fenton, "Preserving Electronic Scholarly Journals: Portico," *Ariadne* 47, http://www.ariadne.ac.uk/issue47/fenton/ (accessed October 15, 2006).

43. Ibid.

44. Stemper and Barribeau, "Perpetual Access," 94.

45. E.L. Bogdanski, "Serials Preservation at a Crossroads," *Serials Review* 32, no. 2 (2006): 71.

46. Google, "Google Book Search Partner Program: An Online Book Marketing and Sales Program," http://books.google.com/googlebooks/publisher.html (accessed August 26, 2006).

47. University of Michigan, "Google, the Khmer Rouge and the Public Good," February 6, 2006, http://www.umich.edu/pres/speeches/060206google.html (accessed August 26, 2006).

48. PubMed, "Central Overview," http://www.pubmedcentral.gov/about/intro .html (accessed August 27, 2006).

49. Geller, "Models for E-journal," 48.

50. B. A.Winters, M. F. Smith, D. LaFrenier, M. Geller, D. Jaeger, T. J. Sanville, and M. P. Fletcher, "Responsibility for Preserving and Archiving Electronic Resources: Whose Job is it Anyway?" *Serials Librarian* 40, no. 3-4 (2001): 419-424.

51. C. Clennon, W. A. Shelburne, and T. H. Teper, "Building Publishers' Journal Archives in the Committee on Institutional Cooperation," *Serials Review* 32, no. 2 (2006): 87-91.

52. Mark Ware, "ALPSP Survey of Librarians on Factors in Journal Cancellation [Summary]," (Worthing,West Sussex: Association of Learned and Professional Society Publishers), http://www.alpsp.org/publications/libraryreport-summary.pdf (accessed April 30, 2006).

53. Kristen Antelman, "Self-Archiving Practice and the Influence of Publisher Policies in the Social Sciences," *Learned Publishing* 19, no. 2 (2006): 85. http:// eprints.rclis.org/archive/00006023/01/antelman_self-archiving.pdf (accessed May 22, 2006).

54. Sally Morris, "'Version Control' of Journal Articles," http://www.niso.org/ committees/Journal_versioning/Morris.pdf (accessed May 20, 2006).

55. James Currall and Peter McKinney, "Investing in Value: A Perspective on Digital Preservation," *D-Lib Magazine* 12, no. 4 (2006), http://mirrored.ukoln.ac .uk/lis-journals/dlib/dlib/dlib/april06/mckinney/04mckinney.html (accessed April 21, 2006).

56. N. F. Foster and S. Gibbons, "Understanding Faculty to Improve Content Recruitment for Institutional Repositories," *D-Lib Magazine* 11, no. 1 (2005), http:// www.dlib.org/dlib/january05/foster/01foster.html (accessed August 13, 2006).

57. International Federation of Library Associations and Institutions, "Preserving the Memory of the World in Perpetuity: A Joint Statement on the Archiving and Preserving of Digital Information," http://www.ifla.org/V/press/ifla-ipa02.htm (accessed June 4, 2006).

58. Erik Oltmans and Adriaan Lemmen, "The e-Depot at the National Library of the Netherlands," *Serials* 19, no. 1 (2006): 63-64. (The list of publishers in the print version of this article is incorrect; the corrected list appears in the online version.)

59. Stemper and Barribeau, "Perpetual Access," 96.

60. Gail Hodge and Evelyn Frangakis, "Digital Preservation and Permanent Access to Scientific Information: The State of the Practice," http://cendi.dtic.mil/ publications/04-3dig_preserv.pdf (accessed April 30, 2006).

61. Rebecca A. Graham, "Evolution of Archiving in the Digital Age," *Serials Review* 26, no. 3 (2000): 59-60.

62. N. Andrews, "The Oxford Journals Online Archives: The Purpose and Practicalities of a Major Print Digitization Program," *Serials Review* 32, no. 2 (2006): 80.

63. Elsevier, "Elsevier and Koninklijke Bibliotheek Finalise Major Archiving Agreement," http://www.elsevier.com/wps/find/authored_newsitem.librarians/ companynews05_00020 (accessed April 25, 2006).

64. JSTOR, "JSTOR: Participating Publishers," http://www.jstor.org/about/part .publishers.html (accessed October 15, 2006).

65. Koninklijke Bibliotheek, "Publishers," http://www.kb.nl/dnp/e-depot/dm/uitgevers-en.html (accessed October 15, 2006).

66. LOCKSS, "Publishers and Titles," http://www.lockss.org/lockss/Publishers_and_Titles (accessed October 15, 2006).

67. Ithaka Harbors, Inc., "Portico: Participating Publishers," http://www.portico.org/about/part_publishers.html (accessed October 15, 2006).

68. National Institutes of Health, "PubMed Central Journals: Full List," http://www.pubmedcentral.nih.gov/fprender.fcgi?cmd=full_view (accessed October 15, 2006).

69. International Federation of Library Associations and Institutions, "Nature and Role of Legal Deposit," http://www.ifla.org/VII/s1/gnl/chap1.htm (accessed April 24, 2006).

70. Llyfrgell Genedlaethol Cymru/The National Library of Wales, "Agency for the Legal Deposit Libraries," http://www.llgc.org.uk/cla/ (accessed April 24, 2006).

71. Department for Culture, Media and Sport, "Libraries & Communities: Legal Deposit," http://www.culture.gov.uk/libraries_and_communities/legal_deposit/ (accessed April 24, 2006).

72. Rick Weiss, "Government Health Researchers Pressed to Share Data at No Charge," Washingtonpost.com, March 10: A17, http://www.washingtonpost.com/wp-dyn/content/article/2006/03/09/AR2006030901960.html (accessed May 25, 2006).

73. Michael Spinella, "The Andrew W. Mellon Foundation: JSTOR," http://www.mellon.org/programs/otheractivities/JSTOR/JSTOR.htm (accessed May 25, 2006).

74. Digital Library Federation, "Archiving Electronic Journals," http://www.diglib.org/preserve/ejp.htm (accessed October 15, 2006).

75. William and Flora Hewlett Foundation, "Grants at the William and Flora Hewlett Foundation," http://www.hewlett.org/HewlettDev/FoundationWideGrantBrowserTemplate.aspx?NRMODE=Published&NRORIGINALURL=%2fGrants%2fgrantBrowser%2ehtm%3fid%3d5056&NRNODEGUID=%7b1EBC7B6F-38FB-472D-9EC9-913D001BAE93%7d&NRCACHEHINT=NoModifyGuest&id=5470 (accessed October 15, 2006).

76. Wellcome Trust, "Wellcome Library: Projects," http://library.wellcome.ac.uk/projects.html (accessed June 8, 2006).

Chapter 4

Case Study:
Evolving Purchasing
and Collection Models for Serials

Hilary Davis

INTRODUCTION

Increased demand of e-resources by patrons and the need for management tools to manage these electronic collections by librarians have spurred many libraries to adjust their collections and acquisitions strategies. This chapter gives an outline of the context in which electronic library collections are evolving and the implications of concomitant changes for collections and acquisitions functions. Each case study includes the background which led the case libraries to embark on a new model for serials collecting and purchasing. Processes and criteria used by each case library are described so that readers can apply some of these methods to work through their own initiatives for building robust electronic collections. Consequent impacts on budgeting, long-term planning, staffing, and workflows are also reviewed as lessons learned for the benefit of other libraries considering experimenting with pay-per-view serials access or converting serials collections to electronic-only format. Evaluation tools and benchmarks used by the case libraries provide valuable understanding of the impact of alternative purchasing models and better prepare librarians for undertaking similar efforts.

King and Tenopir found that approximately 80 percent of use of library resources takes place in electronic format because users find that e-resources are easier and more convenient to use and save more time.[1] To meet users' expectations, libraries that are already functioning within tight resource constraints must develop cost-effective collection strategies.[2] As a result, more and more libraries are shifting their journal collections to electronic-only format and realigning their collection development policies to support this framework. In many cases, these libraries are including these policy changes in very public venues designed to justify their decisions and involve stakeholders, especially faculty and students. These changing collections and policies have far-reaching implications from budgetary allocations to workflow processes, staffing, and relationships with user communities.

Hunter provides an excellent overview of the history of transitioning to electronic serials. She points out that, even though electronic-only subscription models were available from some major publishers since the mid-1990s, the majority of academic libraries did not begin to embrace this subscription model until 2000. Hunter found that libraries based their decisions to "go e-only" on several factors: an inability to support both print and electronic formats; attractive cost-savings for electronic-only subscriptions; and a need for realized savings with regard to shelf space and binding costs.[3]

Many authors have reported on the rising costs of serials and the impact on library budgets. Dingley reports that the average cost of serials increased by 731 percent, from $54 in 1984 to $449 in 2005.[4] These increases have had a dramatic and all too familiar effect on libraries' materials budgets. In his analysis of Association of Research Libraries (ARL) libraries' uptake of e-resources from 1997 to 2004, Stoller finds that, on average, spending for e-resources by libraries has increased by 85 percent each year.[5] This represents a jump from $50.5 million in 1997 to $301.7 million in 2004 amongst a sample of the largest and smallest ARL academic libraries.

Apart from budgetary implications, the uptake of e-resources in academic libraries has been controversial for many reasons. Reservations of librarians regarding the transition to electronic-only collections have centered on concerns over potential negative feedback from users. Will there be a significant difference in quality of service if content is delivered via a networked server compared with traditional print access? What are the logistics of assimilating collections of libraries participating in consortial serials access arrangements? What may be the impact on users of a potential inability to commit continuing funds for electronic access and consequences of publishers discontinuing service? Could there be potential delays in publication time for content delivered online? How will universities handle higher demand on campus networking infrastructure to support online access? Will there be assurances that perpetual access rights can be met and that publishers will take appropriate archival measures?[6] Cole highlights that consistent and perpetual access is an important issue in the ever-changing marketplace of scholarly communication.[7] Changes in journal titles, publishers, and pricing and packaging models require careful thought and planning for a shift to electronic collections and purchasing models. Libraries experimenting with providing access to serials via electronic-only serials collections have reconciled many of these concerns by negotiating consortial agreements for electronic access serials and working with publishers to secure perpetual-access rights. Users have for the most part demonstrated that they expect electronic access to serials, thereby influencing institutional decisions to build more robust networking infrastructures as well as decisions to commit more funds for electronic content.

In contrast, for those libraries that are providing pay-per-view serials access, most of these concerns are irrelevant because they are focusing only on access, not on ownership or rights to content. Pay-per-view serials access is fundamentally different from electronic-only access in that pay-per-view access is not concerned with traditional archival or preservation roles. Nevertheless, as Montgomery points out, moving toward an electronic-only serials collection or merely just access via pay-per-view involves giving up some control and taking on some risk: "Given the chaotic state of scholarly publishing, we do what is best for our current users within the limits of our budget. We understand that this electronic collection based upon publishers' packages is fragile since it is extremely sensitive to budget reductions."[8]

OVERVIEW OF THE CASE LIBRARIES

The libraries discussed in this chapter have implemented two differing strategies for providing electronic access to serials: a wholesale shift to electronic-only journal collections and pay-per-view serials purchasing. The following institutions were selected for case studies based on their efforts to engage in evolving serials collections and purchasing activities and because they represent a cross-section of different types of academic institutions: University of Nevada, Las Vegas, Cornell University, University of North Carolina at Greensboro, and University of Wisconsin-Madison (see Table 4.1).[9] Represented in this small sample are libraries that range from large land-grant institutions to younger liberal-arts institutions. Respondents from each institution were interviewed by phone and e-mail during September 2006. The diversity and similarities in strategies and perspectives of the case libraries provide guidance for libraries that are considering following in their footsteps as well as enable reflection and evaluation

TABLE 4.1. Characteristics of Case Libraries

Library System	Date Founded	Carnegie Classification	Student Population as of Fall 2006	Total Budget	Collections and Acquisitions Budget
University of Nevada-Las Vegas (UNLV)	1957	Large four-year doctoral degree-granting university	28,000	$11.3 million (2004)	$4.6 million (2004)
Cornell University	1865	Large land grant four-year doctoral degree-granting university with very high research activity	20,000	$50.2 million (2004-2005)	$15 million (2004-2005)
University of North Carolina at Greensboro (UNCG)	1891	Large four-year doctoral degree-granting university with high research activity	14,000	$8.6 million (2004-2005)	$3.6 million (2004-2005)
University of Wisconsin-Madison (UW)	1848	Large land grant doctoral degree-granting institution with very high research activity	41,000	$39.2 million (2004)	$10.5 million (2004)

of experiences with other libraries that have already been experimenting with pay-per-view serials access or have committed to electronic-only serials collections.

All four universities are classified as large four-year doctoral degree-granting institutions under the Carnegie Classification system. University of Wisconsin-Madison is the largest in terms of student population with 41,000 students reported for Fall 2006. University of North Carolina at Greensboro is the smallest university with 14,000 students reported for Fall 2006. Founded in 1957, the University of Nevada, Las Vegas (UNLV) is by far the youngest of the four case universities. Cornell University and University of Wisconsin-Madison are both members of the ARL and rank in the top ten and top twenty, respectively. While Cornell University Libraries reported the greatest operating budget ($50.2 million for 2004-2005), University of North Carolina at Greensboro and University of Nevada, Las Vegas report the largest percentage of the operating budget devoted to collections and acquisitions (approximately 40 percent).

CASE STUDIES FOR SHIFTING
TO ELECTRONIC-ONLY SERIALS COLLECTIONS

University of Nevada, Las Vegas (UNLV) Libraries

(Respondents: Xiaoyin Zhang [head, Materials Ordering and Receiving] and Michaelyn Haslam [librarian, Materials Ordering and Receiving])

Background

Since 2000, UNLV Libraries has been incrementally converting print journal subscriptions to electronic-only subscriptions. The motivations for this ongoing shift to electronic-only serials are threefold and mirror the motivations of other research libraries that are transitioning to electronic-only subscriptions. These motivations include,

1. User expectations: to meet the increasing expectations of faculty and students that online access to journals and databases should be extensive and ubiquitous, especially as research becomes a

  remote practice and as distance-education programs continue to develop and mature.

2. User needs: to provide more complete instruction services and resources to satisfy the learning and research needs of the user community.

3. Functional necessity: library staff at UNLV recognized that to more efficiently manage serials workflows and processing, they had to make a decision to collect print or electronic formats, not both.

  User expectations and the need for more efficient and manageable systems coalesced to form a tipping point for shifting to an electronic-only serials collection.

  Before the inception of the shift to electronic-only serials collection practices in 2000, UNLV Libraries had 6,200 print subscriptions and a handful of e-resources. Since then, UNLV Libraries has converted between 200 and 500 print subscriptions each year to an electronic-only format and has added many new electronic serials. As of 2006, the serials collection consists of 17,000 electronic subscriptions and only 2,000 print subscriptions. As UNLV Libraries has worked to convert the collection, policies have been adopted stipulating that any new subscriptions should be electronic-only, when possible.

  What is it about UNLV Libraries that makes this shift to electronic-only serials successful? As most academic libraries will attest, user demand and expectation for electronic access to resources have played a major role in the acceptance of electronic-only serials collections. Since 2000, UNLV Libraries has expanded consortial relationships, bringing in 3,500 new subscriptions through shared licensing and negotiation practices.[10] In addition to changes in collection policies favoring electronic-only journals, the materials budget for electronic-resources has also been increased to support these new policies. As a result, workflows were reorganized and staff resources were reengineered to accommodate these new practices.

Process and Criteria

  In accordance with the criteria that UNLV Libraries uses to build the electronic-only serials collection, electronic-only subscriptions are favored with few exceptions when both print and electronic formats are available. Print subscriptions are maintained for Special Collection acquisitions, when print is made mandatory with electronic format by publishers and/or vendors, and in cases where a password is required for electronic access.

Impacts on Budget and Long-Term Planning

Print journals and monographs are being cut to support growing e-resource needs and the tools to support them. At UNLV Libraries, the shift to electronic-only serials has meant that spending on print serials—subscription costs as well as nonsubscription costs (e.g., staffing, materials, binding, shelving, space)—has been cut back and will continue to be cut back. According to the respondents, while electronic subscriptions usually do not cost more than print-only subscriptions, the impact of inflation on "buying power" remains the same. Incentives such as buying in bulk via e-journal packages have fueled the fire of electronic-only journal collection-building. Discounts for bundled packages and journal archives or backfiles are just as popular at UNLV Libraries as they are elsewhere. Owing to these incentives, funds allocated for books at UNLV Libraries have been used more and more to support journal subscriptions; this trend, of course, is not unique to UNLV Libraries. Moreover, with a heavy focus on e-journal collections, UNLV Libraries has had to add new budget lines to purchase tools for managing access such as e-journal finders, link resolvers, and metasearch products. The respondents also described a recognized need to increase staffing resources to better analyze and use statistics and handle complex licensing.

One major concern with respect to long-term planning and the future of the collection for the respondents is that UNLV Libraries is leasing much of the journal content rather than owning the content. Backfile or journal archive content, while "owned," is hosted remotely, leaving the onus of archiving and access on the publisher or vendor and taking control away from libraries.

The shift to electronic-only serials has also impacted long-term strategic planning at UNLV Libraries. In the Strategic Plan for 2005-2010 (www.library.unlv.edu/about/strategic_goals.pdf), UNLV Libraries has stipulated that increases in the base budget and any one-time end-of-year funds will be used to enhance the availability of resources, especially e-resources. Any additional future funds will also be used to organize an evaluation of all remaining print subscriptions and explore the possibility of leveraging aggregator databases as alternative means of access to low-use print resources.

Impacts on Staffing and Workflows

UNLV Libraries' transition from print to e-resources has had major ramifications for job responsibilities and position descriptions as well as organizational structure. In the fall of 2000, the unit primarily responsible for print serials processing and management was reorganized so that members of that unit could assume the responsibility of processing and ongoing maintenance of all e-resources. Workflows for selecting, acquiring, receiving, and claiming were reconfigured. All position descriptions were altered to reflect the organizational and functional changes to the unit. Serials receiving processes are now handled by a student assistant working twenty hours a week. Claims processes, which are relevant only for print journals, have been reduced. In addition, a new position, electronic resources librarian, was created in 2003 to oversee the management and processing of e-resources. Down the road, UNLV Libraries anticipates that processing of print serials will be established as a separate functional unit.

Evaluation Tools and Benchmarking Methods

The UNLV Libraries' respondents explained that it is too soon to gauge the economic impact and efficiency of workflows implemented as a result of the shift to an electronic-only serials collection. Too many changes and fast expansion to electronic-only collection practices have made it difficult to measure outcomes. Nevertheless, a combination of in-house and vendor-developed tools has been used to manage the transition and will continue to be used on a long-term basis. Commonly used programs such as Microsoft Access and Excel as well as the integrated library system, Innovative Interfaces, Inc. have supported most of their needs.

Cornell University Libraries

(Respondents: John Saylor [director, Engineering Library and Collection Development for National Science Digital Library] and Bill Kara [head, Technical Services, Mann Library])

Background

According to the respondents, two years ago, Cornell University Libraries investigated potential subscription and nonsubscription savings (e.g., staff, resources, space) of converting the serials collection to electronic-only format. The respondents explained that dramatic savings could be attained by eliminating duplicate subscriptions. Staff at Cornell University Libraries set forth a goal of cancelling duplicate serials and converting existing subscriptions to electronic-only access for 4,000 journal titles over a three-year period. The titles spanned the sciences, social sciences, and humanities. In the life sciences, many duplicate subscriptions existed across different Cornell University Libraries—these "low-hanging fruit" were the first to be cancelled. Thus far, the library system has incurred 1,500 subscription changes from print to electronic format with around 1,000 more being converted to electronic-only for 2007. Much of the savings realized thus far have resulted from the elimination of duplicate print subscriptions amongst the life sciences. The transition of engineering, mathematics, and physical sciences journals that have already transitioned to electronic-only subscriptions has received strong, positive feedback from faculty and students in these disciplines. The latest push for the transition from print to electronic has moved beyond science libraries to focus more on humanities and social sciences journals.

Process and Criteria

Decision-making criteria for converting journals to electronic-only or maintaining print are described on the Cornell University Libraries Web site on scholarly communication (http://www.library.cornell.edu/scholarlycomm/serials/eonly). The criteria are based on function within the context of a particular institution (e.g., high profile journals, importance for browsing, image quality, aesthetic value), availability of electronic archives in accordance with publishers' policies for long-term preservation, commitment to preservation of journals at a local level, responsibility to retain print subscriptions at a local level (e.g., based on consortial agreements), time delays and reliability of publication based on publishers' practices, and quality/quantity of electronic versus print content based on publishers' practices. This is an excellent set of criteria for other academic libraries to consider in planning a conversion to an electronic-only serials collection. Some of these criteria are library-specific and some are publisher-specific. For example, considerations of the artifactual or aesthetic value of serials will certainly differ from one library collection to another while guaranteed archival access to serial content is dependent on publishers. Other criteria might include the presence or absence of complete holdings for serials and the ownership of online serials backfiles.

Impacts on Budget and Long-Term Planning

In terms of budgetary savings, the respondents from Cornell University Libraries reported an initial savings of 10 percent by eliminating duplicate subscriptions and converting many existing subscriptions to electronic-only. In terms of long-term planning, the respondents indicated that the project is still in flux—too many aspects are still evolving in terms of statistics collecting, package plans, title changes, and ownership changes. In addition to the need for new systems and technical reporting functionality to support an electronic-only serials collection, administrators at Cornell University Libraries are considering journal use and necessity of physical presence of print journals as they plan for the future. Job descriptions are also changing and new positions are being considered for the future.

Impacts on Staffing and Workflows

The respondents indicate that there is anecdotal evidence that reallocation of staff time and responsibilities have had a positive impact. Even so, vacant positions have been realigned or recouped in different ways, and many staff responsibilities have changed regarding record maintenance and purchase order workflows.

By undergoing the process of reviewing titles for conversion to electronic-only format, librarians and staff were able to get a better handle on serials records and payments and developed better solutions for managing subscriptions. This experience was also an opportunity to bring more librarians up to speed on evaluating statistics on use of the collection.

Evaluation Tools and Benchmarking Methods

The type of information collected to evaluate the effectiveness of converting the collection to electronic-only format included subscription savings for duplicates, subscription savings for electronic-only acquisitions, number of issues no longer received, number of journals no longer needing to be bound, and the number of check-ins pre- and posttransition to an electronic-only serials collection. Librarians involved in the conversion process are also estimating the time required to review the physical quality of journals, review licenses, and update licenses for different kinds of publishers to better understand how to approach these processes more efficiently. Some publishers have hundreds of titles, so they can all be dealt with easily, while others have only a few titles. Different scales have dramatic implications for time spent on decision making.

CASE STUDIES FOR PAY-PER-VIEW SERIALS ACCESS

University of North Carolina at Greensboro (UNCG) Libraries

(Respondent: Beth Bernhardt [e-journals/document delivery librarian and head of Interlibrary Loan, ILL])

Background

Several major changes across academic programs at the University of North Carolina at Greensboro brought about a project at UNCG Libraries to provide access to serials via a pay-per-view model. In 2001, a distance-education program was established with its own source of funding. Since that time, UNCG Libraries has provided access for the distance-education students and faculty. For the most part, ILL is insufficient for these users, who expect quick and efficient access to resources. In addition, UNCG Libraries has moved from a focus on music education to expanding its programs to include a new science center specializing in biotechnology and genetics. As a result, UNCG Libraries has had to determine how to provide access to subject-specific journals that best serve these new programs.

When these developments were taking place, EBSCO and First-Search began offering article pay-per-view access services. UNCG Libraries decided to try both services, effectively adding many new journal titles to the collection. Over the past few years, they have been working with Blackwell, Kluwer, Ingenta, Elsevier, Wiley, Ovid, and the American Institute of Physics to provide pay-per-view serials access.[11] Overall, UNCG Libraries has added 3,000 to 3,500 journals via pay-per-view serials access.

Process and Criteria

Journals included in the pay-per-view serials access model are those journals that are not already owned by UNCG Libraries, have a surcharge for online access, or have an imposed embargo period. In some cases, access to newly requested journals is provided on a provisional basis via pay-per-view access to determine actual use before committing continuing funds to the new journals.

After one year of experimenting with pay-per-view serials access, the Libraries recommended that twenty-one titles be added to the collection because it was more cost-effective to subscribe to the journals than to continue to provide access via a pay-per-view mechanism. Most of these subscriptions were for psychology journals as well as social science journals. The science journals still proved too cost-prohibitive to subscribe to, so the Libraries retained pay-per-view access for these titles.

Impacts on Budget and Long-Term Planning

At the project's inception, UNCG Libraries received funding for the pay-per-view serials access project from the distance-education program. Using pay-per-view serials access as an alternative to subscriptions, UNCG Libraries paid an average of $400 to $500 per year per journal.

As evaluation and long-term planning continue, UNCG Libraries is trying to find the best way to provide access to students and faculty. As some recently established consortial partnerships provide much more cost-effective journal access, some of the pay-per-view serials access models are being phased out. Currently, UNCG Libraries continues to receive funds from the distance-education program to support pay-per-view access for many science journals.

Impacts on Staffing and Workflows

The respondent described the overall impact of the pay-per-view serials access approach as minimal in terms of hiring more staff. Much of the additional work for the pay-per-view project was appended to an existing position responsible for serials management, with some of the responsibilities for collecting statistics being shared with another staff position. With regard to access and copyright issues decisions, no additional work procedures were necessary. As patrons can purchase and download articles on their own via the pay-per-view serials access model, the responsibility of UNCG Libraries to follow proper copyright restrictions is removed. Users are made aware of copyright issues when they download the articles directly from the publishers.

Evaluation Tools and Benchmarking Methods

As part of their evaluation process, staff members at UNCG Libraries have tracked statistics including dates of access, dates of publications accessed, titles of articles, number of articles purchased, cost of each article, and the status of the journals accessed if owned in print. Most of this data are provided by the publisher or vendor along with invoices or regular reports.

To enable users to discover the content provided via pay-per-view serials access, UNCG Libraries added these journal titles to its Journal Finder tool. Each user must log into the network and authenticate in order to access these journals. Under this setup, UNCG Libraries can also track usage and requests deriving from both on-campus and off-campus users.

University of Wisconsin-Madison (UW) Libraries

(Respondents: Deborah Helman [director, Wendt Library], Richard Reeb [associate director, Collection Development and Technical Services], and Jean Gilbertson [director, Steenbock Library])

Background

Over the past several years, the UW Libraries has been trying to control the amount of acquisitions and collections money given to a few large publishers. Initially this was accomplished by reducing the number of duplicate print subscriptions and by cancelling unique high-cost, low-use subscriptions. With the support of the director of libraries and campus administration, librarians eventually wanted to determine if they could greatly improve user access to cancelled journals by buying articles directly from publishers rather than relying on ILL channels. The assumption was that purchasing articles would cost significantly less than the subscription price. The pilot project was initiated in January 2005 and, while evaluation of the project is ongoing, UW Libraries has reached the point where pay-per-view is proving to be a cost-effective and user-friendly alternative for many high-cost journal subscriptions with relatively low demand.

Process and Criteria

The journals that were included in the project were cancelled over a period of several years. Therefore, the criteria used to select the journals carried over from prior serials cancellations: subscription cost, inflation rate, usage levels, relevance of subject content, faculty/researcher feedback, and alternate routes for providing access. The project currently includes 700 previously cancelled journal titles. Preliminary data showed that about 60 percent of the cancelled journals generated five or less article requests in the first year. There were 5,300 articles purchased for users via pay-per-view in the last fiscal year.

After the first year of the project, a few titles were identified for which subscriptions should be reinstated because article demand exceeded or met subscription cost. The respondents noted that, since social science journals tend to be less expensive than science journals, using pay-pre-view for social science journals could more quickly reach the cost of a subscription.

Impacts on Budget and Long-Term Planning

Funds for the project did not come from the collection budget, but came from funds secured by the director of libraries. One respondent made an important distinction in the approach to the project: "this project is being supported through special funding from campus for journal 'access' rather than journal 'acquisitions.'" Partly through the bulk purchase of article "tokens" and also through negotiating a license agreement, UW Libraries has been able to reduce the per-article cost significantly.

Impacts on Staffing and Workflows

UW Libraries has not explored the pay-per-view serials access model enough to determine its potential impact on long-term collection development policies. However, it is clear that the pay-per-view option will impact decisions about serials renewals. Recent years' budgets have been so restrictive that UW Libraries is continually forced to re-think collection development strategies. This project has been part of this continuum of assessment, evaluation, and response to increased journal costs and inadequate budgets. Librarians at UW Libraries have used this as an opportunity to spend money differently and focus spending on certain core collections.

All of the pay-per-view rapid article delivery processing is conducted by the campus ILL staff. As a result of the project, ILL staff members have changed their workflows significantly. In the more streamlined process, each staff member follows each request through to delivery, decreasing transaction times by ordering articles directly from publishers and simplifying the number of steps necessary compared with a typical ILL transaction.

Evaluation Tools and Benchmarking Methods

A feedback and marketing mechanism for the project was established by including an e-mail to each article requestor that indicated that their article access was part of a project to improve access to journals no longer available on campus (while not revealing all journal titles that were included in the project). The respondents indicate that UW Libraries is still evaluating if the "increased direct costs (for article buying rather than using ILL) and the indirect staffing costs (which are harder to measure) are justified by what has turned out to be an overwhelmingly positive response by users."

CONCLUSION

The case studies presented in this chapter offer differing perspectives for a nearly wholesale shift to an electronic-only journal collection and an emerging method of access via a pay-per-view serials purchasing model. Librarians from a cross-section of different types of academic institutions were interviewed regarding their experiences and perspectives with evolving purchasing and collection models for serials.

University of Nevada, Las Vegas Libraries and Cornell University Libraries were interviewed regarding their experiences transitioning to a nearly wholesale electronic-only serials collection. Their experiences vary across issues of staffing, technical infrastructure, and budget. For both institutions, ensuring that all staff are equally well informed and have a common understanding of goals and issues related to the transition to electronic-only serials has been a challenge. In terms of staffing, UNLV Libraries found that more staff time has been necessary to maintain and resolve access-related issues than was initially expected. Hesitations amongst staff and users at both institutions have had to be overcome. At Cornell University Libraries, some selectors have been more cautious about this transition while others have embraced the move to electronic-only access. In the humanities, the option to go electronic-only has been challenged by faculty due to concerns about archival access and the preservation policies of publishers. Many of the subsequent conversations between faculty and library staff have revolved around scholarly communication issues and have resulted in an opportunity to educate both librarians and faculty about these issues.

Overall, the experiences at both institutions have brought about a common awareness of collection issues amongst librarians, faculty, and students. The respondents at Cornell Libraries explained that the experience has been unifying for selectors because it has given them the opportunity to work together on a common project. The UNLV Libraries' respondents underscore the importance of establishing an internal system of open and clear communication as well as support for staff who are expected to adapt to changes in workflows. Alongside a solid infrastructure for communication and support, they highlight a critical need for a good tracking mechanism for all electronic-journals to enable uninterrupted service.

The nonsubscription savings at Cornell Libraries have had a very important impact on the system; they have been able to clean up records, check on payments, identify ways to better manage subscriptions, and engage more selectors in taking an active role in evaluating use statistics. They have also given feedback to publishers and enabled them to provide better services. One of the overall lessons learned was that there were many instances where collections and purchasing functions need more centralized procedures for gathering, researching, and reviewing statistics about the use and value of the collection as a whole.

The respondents at UNLV Libraries also point out that users have been very happy with the efforts of UNLV Libraries to make as many journals as possible available electronically. The respondents comment that, "Usually the electronic use is so much higher than for the print equivalent, it feels like we are providing a much needed service to our library community. It is worth working through rough spots in order to provide access to resources that get used so much."

Librarians at the University of North Carolina at Greensboro Libraries and UW Libraries were interviewed regarding their experience with a pay-per-view model of serials access. Overall, for both institutions, it is evident that pay-per-view serials access is different for each library in terms of subject areas that are targeted for pay-per-view access and impacts on workflows and staffing.

Reactions from administration at both institutions have been very supportive of these experimental methods devoted to saving money and time. Feedback from users has been positive with few exceptions. At UNCG Libraries, some users complained that some of the publishers' interfaces are not intuitive with regard to ordering articles via the pay-per-view system. Feedback at UW Libraries has also been positive with a turnaround time averaging between a few hours to less than half a day. While enduring the pressure of journal cancellations, faculty and researchers are encouraged when offered the rapid article delivery access. In particular, users requesting articles under grant-proposal and manuscript deadlines "greatly appreciated" the rapid article delivery access. The UW respondents added that "many users commented on how the rapid delivery improved their productivity as researchers (being able to move projects ahead more quickly)." Some users made suggestions about how to spend library funds or expressed concern about spending money on such a granular scale. Issues of subscription cost for a journal versus cost per article opened opportunities for educating and communicating with faculty about the costs of scholarly communication. In addition, the quality of images, especially in the sciences, is often decreased after going through the standard photocopying and scanning process of ILL. As the articles come directly from the publisher, pay-per-view serials access has been a great benefit for researchers needing high-quality images. The respondent at UNCG Libraries added that libraries should rethink a pay-per-view serials access strategy if the collections and acquisitions budget becomes overused. Every library should already have a core set of journals in their collection; pay-per-view articles should only meet the needs for journals at the outskirts of the core collection.

For UNCG Libraries, taking advantage of consortial leveraging for subscription costs has eliminated one of the major disadvantages of pay-per-view serials access: the lack of archival access rights. Since UNCG Libraries has been participating in the Carolina Consortium, much of the pay-per-view serials access has decreased (approximately 80 percent). With regard to their consortial relationships in which access to serials is dependent on all partner libraries participating in a publisher license, both institutions are considering the impacts on their partner libraries in choosing to engage in pay-per-view serials access in terms of not having as many journals to share with consortium partners.

Regarding publishers' reactions, the respondent at UNCG Libraries reported that some publishers are leery to get involved with pay-per-view serials access because of concerns that it might have a negative impact on their subscription revenues. In addition, some publishers were new to the idea of pay-per-view serials access and had to set up new processes for accounting and authentication/recognition in order to accommodate the needs of UNCG Libraries. Both institutions have discovered that there is a high level of record-keeping necessary to document and evaluate a project like this, including tracking articles being purchased, journal titles, subscription costs, and costs of articles among other details.

Wholesale conversion to electronic-only serials collections and pay-per-view serials models do have interesting implications for library collections and acquisitions functions as well as future collection development strategies. By experimenting with evolving collections and purchasing models, the four case libraries have developed core serials collections and have looked more closely at how and why libraries subscribe to titles and how selection decisions impact future decisions regarding collection focus, preservation, and access. With regard to impacts on collection focus, consortial arrangements with publishers for electronic-only collections may be an attractive option but it comes with its own drawback, namely loss of specialized collections. Eells warns against the homogenization of library collections and suggests that "as decisions are made on how to configure collections, and on how to distribute limited library funding, librarians must maintain a strong connection to their users and an awareness of their preferences and needs."[12]

Opportunities to connect with users and develop an awareness of their expectations have arisen out of the experiences of the case libraries. As demonstrated by the Cornell University Libraries' experience, concerns and misunderstanding about the overall impact of evolving purchasing and collection models on the economics of scholarly communication can be addressed with librarians, faculty, and administrators.

Neither model is a perfect option for most libraries. Customization of processes, uncovering roadblocks, dealing with resistance to change internally and externally, and confronting concerns about an uncertain future will vary across different library contexts. However, both models offer opportunities to meet the increased demand of e-resources by patrons and develop tools and strategies to manage electronic collections.

NOTES

1. Donald W. King and Carol Tenopir, "An Evidence-Based Assessment of the 'Author Pays' Model," *Nature Forum* (2004): http://www.nature.com/nature/focus/accessdebate/26.html (accessed January 12, 2007).

2. Peter Boyce, Donald W. King, Carol Montgomery, and Carol Tenopir, "How Electronic Journals are Changing Patterns of Use," *Serials Librarian* 46 (2004): 121-141.

3. Karen Hunter, "Going 'Electronic-Only': Early Experiences and Issues," *Journal of Library Administration* 35 (2002): 51-65.

4. Brenda Dingley, "U.S. Periodical Prices—2005," http://www.ala.org/ala/alctscontent/alctspubsbucket/alctsresources/general/periodicalsindex/05USPPI.pdf (accessed January 12, 2007).

5. Michael Stoller, "A Decade of ARL Collection Development: A Look at the Data," *Collection Building* 25 (2006): 45-51.

6. Hunter, "Going 'Electronic-Only,'" 61-63.

7. Louise Cole, "A Journey into E-resource Administration Hell," Serials Librarian 49 (2005): 141-154.

8. Carol Hansen Montgomery, "The Evolving Electronic Journals Collection at Drexel University," Science and Technology Libraries 24 (2003): 184.

9. Xiaoyin Zhang and Michaelyn Haslam, e-mail exchange with author, September 21, 2006; John Saylor and Bill Kara, telephone interview with author, September 8, 2006; Beth Bernhardt, telephone interview with author, September 1, 2006; Deborah Helman, Richard Reeb and Jean Gilberston, telephone interview with author, September 18, 2006.

10. Xiaoyin Zhang and Michaelyn Haslam, "Movement Toward a Predominantly Electronic Journal Collection," *Library Hi Tech* 23 (2005): 82-89.

11. Beth Bernhardt, "Pay Per View: A Library's Perspective," Presented at the North Carolina Serials Conference, Chapel Hill, NC, April 2-4, 2003.

12. Linda L. Eells, "For Better or For Worse: The Joys and Woes of E-journals," *Science and Technology Libraries* 25 (2004): 46.

PART II:
EVOLVING STAFF
AND PARTNERSHIPS

Chapter 5

Collaborative Library-Wide Partnerships: Managing E-Resources Through Learning and Adaptation

Joan Conger
Bonnie Tijerina

INTRODUCTION

The transition from print to electronic journals and resources has meant great change for libraries. Users are accessing information in previously unimagined ways, and information producers are creating novel distribution methods. New technologies, a constant influx of new resources, and evolving user expectations make managing e-resources more complex and less linear than managing print resources. Influenced by these shifts, the roles and responsibilities of librarians and library staff are changing. E-resources have called into question routines, processes, policies, and procedures that have been successful for decades and adjustments have not necessarily been smooth. As a result, individuals at all levels in libraries must develop innovative ways to work toward the common mission of providing quality library services for patrons.

This chapter outlines the use of information, communication, and adaptive learning to collaboratively respond to the constant change inherent in e-resource management. Evolving a print collection to include e-resources can dramatically modify an organization's cultural perspectives on workflow, communication, and ideal outcomes. In this chapter, these concepts are illustrated within the context of a collaborative workplace. Skills that individuals and groups can develop to facilitate collaboration and analyze the realistic roles that library personnel from the director to staff can cultivate in order to demonstrate the value each person contributes in a collaborative environment are discussed. Above all, this chapter shows how collaborative working groups and personalized library workflows based on adaptive learning create a more robust and agile decision-making environment.

THE EVOLUTION FROM PRINT TO ELECTRONIC: A NEW WORKFLOW AROUND COMMUNICATION

Today's libraries are in the information delivery business, and anything that slows this task distances the library from successful service to its community of users. Reflecting the goals of today's libraries, librarians must have practices, plans, and policies that will allow them to manage the constant flux of e-resources. They must stay on top of the latest technologies and products while still fulfilling existing responsibilities. Furthermore, librarians must maintain an awareness of the changing needs and habits of users so that libraries do not lose them as customers. No one person can accomplish this alone.

Through collaboration, libraries rely more on a responsive network of professionals with diverse skills and knowledge and less on rules and procedures that can be slow to reflect everyday reality. While rules and procedures are important stabilizing forces, when they cease to keep pace with new information from a wide range of individual experiences, they cease to respond to change. This slow response begins to warp the library's ability for effective service. Indeed, effective service to the patron flourishes in a collaborative environment where informed decision making is paramount, information flows openly and freely, and people are allowed to innovate, learn from mistakes, and question the status quo.

The Linear Print Workflow

To achieve this model for effective service, it is necessary to understand the differences between linear print and nonlinear e-resource workflows. Figure 5.1 shows typical library departments that are involved in acquiring a print resource, such as a journal. Only departments, not specific tasks, are listed. (For an additional outline of print and electronic workflows please see Appendix B of the ERMI [Electronic Resource Management Initiative] Report created by the Digital Library Federation.)[1] The example illustrated in Figure 5.1 is for a midsized library or smaller. Larger libraries will likely distribute the tasks over more work groups.

This example of a typical print journal workflow illustrates the linear process involved in obtaining a journal and moving it from the information provider or producer through library processes to its final destination, access by the user. First, a journal exists that the library would like to purchase. Information about the journal moves from those who approve the purchase, to those who actually purchase it, to those who physically prepare the journal, and finally to those who move it to the shelf. The procedure is repeated, with minor exceptions,

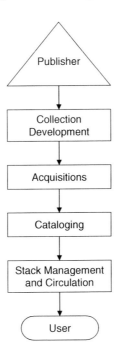

FIGURE 5.1. Print Journals Linear Process

for all journals ordered. The linear nature of the process allows print items to move efficiently from producer through library departments to users, and typically does not require high levels of cross-departmental coordination or communication. The procedure is well known to personnel and information needed to process the item is relatively fixed on the item itself or recorded in a database without expectation of too much change.

The Dynamic E-Resource Workflow

Figure 5.2 shows the analogous acquisition and management of an electronic serial, including which departments might be involved in the evaluation, ordering, and maintenance. As with the previous figure, Figure 5.2 is for a midsized library or smaller. Larger libraries will likely distribute the tasks over more complex communication networks.

The arrows in the figure show an example of who would need to communicate with whom during the management of an e-resource. A publisher or content supplier may be in contact with both the collection development and acquisitions departments as an e-resource's contract and price are negotiated. Collection development may make contact with subject librarians, reference staff, or other public services

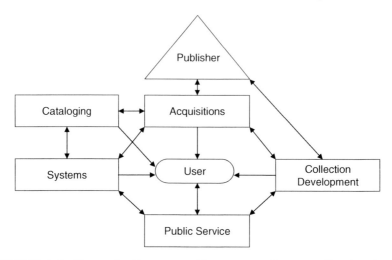

FIGURE 5.2. Electronic Resources Management Process. Reprinted with permission.

professionals about the e-resource's fit within the library's collection, not only in terms of content but also in terms of technological requirements for delivery. Depending on the needs of a product, the acquisitions department might communicate with the collection development, systems, and cataloging departments about special delivery needs. Further, a change in any of these circumstances will require additional awareness and communication during the e-resource's life cycle.

This shift from a linear print workflow to a networked e-resource workflow also leads to the dissolution of clear role delineations. For example, the reference librarian may directly negotiate contracts while the acquisitions librarian may develop a technology expertise. Consolidation and differentiation of roles becomes more fluid as library professionals strive to meet new challenges in a world of constant change.

As a consequence of these changes, the number of people involved in the management of e-resources can become greater and more diffused than the linear workflows required to manage print materials. The flow of information and processes no longer runs in just one direction, from input to output, from placing the order to mailroom to shelf. The attributes of e-resources are now variable rather than fixed, including pricing models, license restrictions, mode of electronic access, special technological considerations, and diverse applications for patron research needs. This variability means that library professionals more often determine on a product-by-product basis what is needed to finally deliver each product to the patron. Procedure ceases to be as fixed and becomes more case-by-case.

For example, the approval process often delays access to the e-resource for significant lengths of time. Access models are often unique to a provider or specific product. The following variations may exist: the material may be owned or leased, providers may have different requirements for simultaneous user limitations or security, and the quality of the coverage years within the content may vary. Before signing a license, the terms may need to be reviewed by more than one department (e.g., use terms by public service professionals and technological requirements by systems professionals) and at diverse levels of administration—indeed, the bundling of e-resources means higher pricing levels and more budgetary oversight. Electronic delivery requirements must be negotiated among several groups or departments to decide where end user access will originate—the online catalog, an A-Z list, and/or a newly created virtual space that supports the product's unique characteristics. Not only must more people be a part of the process of deciding the worth or fit of an e-resource in the library's collection, but changes to any element of the product's purchase agreement can affect any number of related resource requirements. Even the decision to purchase can become drawn out and complex, requiring the input of professionals from across the spectrum of library services.

The complexity of e-resource management not only requires efficient management within our systems, but, if it is to have any validity, it must also create an effective service for the patron. At each step in the process of acquiring and delivering an e-resource, those shepherding the product through the library's processes must consider the user's needs and the product's capabilities. Ultimately, static print workflows are dramatically different from e-resource workflows and cannot accommodate the extreme variation and complexity inherent in the life cycle of e-resources; therefore, policies and procedures must evolve and grow dynamically for libraries to effectively provide access to these resources.

FROM MATERIALS FLOW TO INFORMATION FLOW: SURVIVING CONSTANT CHANGE

The nature of this difference between linear print resource management and networked e-resource management lies in the nature of what the workflows are managing. With print resources, libraries' workflows were oriented toward a physical object that was ordered, received, and shelved. With e-resources, the product itself is not tangible, and workflows center around information about the product. Indeed, a license, a purchase order, and a URL pointing to data stored on a producer's server are all examples of how factors that shape e-resource workflows are oriented toward information about a product rather than a tangible object. Owing to this difference, print workflows and e-resource workflows require entirely different organizational principles.

The management of print resources thrives in a world of clear policy and consistent procedure. One person can perform his or her roles without having to understand or discuss the roles of people several steps down the path of the workflow. As long as each individual does his or her job well, the procedure runs smoothly. The desired outcome is efficiency and consistency for the library and, accordingly, effective use by the library patron.

In contrast, the ERMI workflow illustrated in Appendix B of the ERMI report cited in previous text demonstrates the fluidity of information gathering, communication, and coordination required to manage the resources efficiently within library processes and to make e-resources effective for users. The varied characteristics and requirements of electronic products overwhelm linear, insulated procedural rules of task division and coordination.

This complexity is best managed when decision makers use adaptive learning to balance several conceptual approaches to organizational behavior:

- tapping pools of individual experience for rich input into decision making;
- drawing upon the benefit of understanding information contextually;
- searching for consensus to support collaboration.

Balancing these approaches allows the dynamic processes of e-resource management to flourish without overwhelming library professionals with unwieldy procedures that distance them from their ultimate goal: providing excellent service to patrons. The following few sections describe how active learning gives a library the structure it needs to tap pools of information that exist within stakeholders' experiences, provide an explanation of contextual information, and discuss the decision-making processes utilized by many libraries with particular emphasis on consensus through collaboration. Each of these conceptual approaches provides part of the framework necessary to incorporate adaptive learning into the library environment.

Utilizing Information Pools and Stakeholder Experiences

For stability, linear procedures must remain insulated from all but the most insistent voices for change. Therefore, linear procedures cannot serve the dynamic processes of e-resource management. Dynamic processes must have a constant flow of information to remain viable, flexible, and adaptive to the constant change around them. This information pool represents diverse experiences from all involved and must be organized within a structure that can contain and process it for effective decision making to avoid overwhelming the e-resource management process. The structure for synthesizing these information pools is adaptive learning.

Learning organizations view success and failure very differently from traditional, command organizations. Workers in command organizations receive decisions about tasks from authority figures, procedural policies, cultural norms, and other external sources. In contrast, workers in learning organizations utilize experimentation, evaluation, and learning in order to adapt their work processes to the constantly changing world that surrounds them. In the first instance, decisions are stable and safe, but the outcomes are often disconnected from the evolving expectations of library patrons. In the second instance, decisions are fluid and an organizational acceptance of experimentation and learning means that failure, far from being catastrophic, is eagerly accepted and utilized as a learning opportunity.

Peter Senge writes about traditional organizations getting stuck in "impression management."[2] Decision makers have power when they hold knowledge no one else has, and this power is most often used to protect familiar worldviews from challenge. In these organizations, professionals are evaluated by how well they meet traditional performance measures, and mistakes are seen as failure rather than as stepping stones toward innovation.

Librarians who keep up with new concepts, gain knowledge from their colleagues and environment, and accept that they will be lifetime learners, will improve their organizations, their situations at work, and their passion for their profession.[3] On the job, a genuine commitment to continual learning at both the individual and institutional level is critical to reducing repeated mistakes and fostering the ability to embrace new paradigms.[4]

A linear work process relies heavily on knowledge accumulated in the past. Applying linear processes to e-resources that require dynamic workflows will result in the inefficient handling of e-resources and limit access to the resources for patrons. In addition, professionals in the library environment who are in a constant state of experimenting, evaluating, and learning like vendors, patrons, and competing information providers such as Google and Amazon, will remain disconnected from these linear processes. Defending established work habits and worldviews from adaptation creates isolation and a widening gap between effort and effectiveness. The resulting additional work of defending one's position comes with an attendant anxiety and stress. This rigidity will also retard the innovation of colleagues, hampering mentoring opportunities and growth of the profession as a whole, while negatively impacting the experiences of patrons.

On an organization-wide perspective, traditional organizations often support the linear work process and discourage innovation and risk-taking while learning organizations allow for fluid processes and value staff and stakeholder involvement. Table 5.1, based on

TABLE 5.1. Traditional versus Learning Organizations

Perspectives	Traditional Organization	Learning Organization
Purpose of training	To learn your job	To learn your job, about your organization, and learn how to learn and think about improvements to your job
Strategic intent	To serve current user	To understand and serve current users and think about new or future patrons
Investments	In resources and processes	In people as assets
Decision making	Based on closed systems using Command, Consult, and Vote	Based on open systems using Consensus
Risks	Avoided	Taken and seen as part of learning process
Structure	Hierarchical, not often open to challenge	Adaptive, open to new leaders within the organization
Users	Needs determined (guessed) by the organization	Involved in helping the organization understand their needs
Staff members	Are tools to complete the process, excluded from learning	Are central to learning
Failures	Punished, recorded in evaluations	Ways to learn something new
Successes	Individual rewarded	Whole organization rewarded
Organization focus	On doing	On learning how to do, then doing

Carson et al. (1997), compares perspectives on learning of the traditional organization with the learning organization.[5]

Those who work in learning organizations are always adopting new concepts and unlearning old habits or perceptions. In the current environment, organizations need to work toward becoming learning organizations. Senge advocates transforming these traditional organizations into true forums for open learning, critical thinking, and informed debate; they should move from "impression management" to "collegial management."[6] This collaborative environment of learning professionals—at all levels—facilitates informed decisions in a changing context and, accordingly, is highly compatible with the dynamic nature of e-resource management.

The Pooling of Experiential Information

Information crucial to the successful management of e-resources flows along the spectrum from information creators to information seekers and pools in the experiences of each of these groups. Understanding this flow helps library decision makers obtain information from all stakeholders while also advancing decision makers' interests in the efficiency and efficacy of the e-resource workflow. Figure 5.3

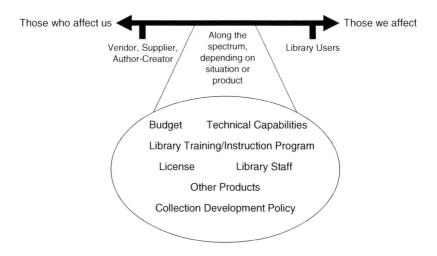

FIGURE 5.3. Influences on Information Pooling. Reprinted with permission

depicts how the influences of different stakeholders range from affecting to being affected by the management of an e-resource.

This figure illustrates the unique and critical role libraries often play between the suppliers of information and those who need to acquire it. This role has become more complicated but also more critical in a digital environment. In a linear print process, the library receives the physical product at the beginning and provides the product for the user at the other end. A clear procedure outlines the decisions of the worker with the tasks for handling a physical object and any creativity lies in the details.

In the dynamic process of e-resource management, decisions depend much more heavily on the relation of the product within a web of experiences. The following examples reflect some of these experiences:

1. the library's experiences of vendor decisions about product creation, sale, and delivery;
2. the experiences of library professionals trying to integrate new e-resources within their library's existing infrastructure of e-resource management tools and workflows;
3. the experiences of the library users as they attempt to incorporate the library's offerings into their own workflows.

An understanding of these pools of experiences is crucial to effective decision making. Products whose design involved the library and the end user are noticeably easier to incorporate into library services. E-resource management decisions that include the perspectives of a diverse range of library staff are less likely to derail on unforeseen resource restrictions. E-resource delivery decisions that include the experience of patrons are more likely to effectively serve their needs.

Whereas a linear process relies on the efficient completion of clear tasks, a dynamic process requires effective incorporation of information from a rich array of sources. Indeed, in e-resource management, the sources of the most accurate and current information are not policies that can go stale over time. Instead, the sources of accurate and current information are stored in the experiences of vendor representatives and library professionals. With this newfound understanding that decisions are dynamic and evolve with the passage of time comes two key realizations: first, that information has its best meaning in context and, second, that one should shift from centralized decisions obtained from one policy or person to enriched decisions made through collaboration.

Contextual Information: The Only Truth in a Dynamic Process

Since decisions are best understood in context, it is important to define contextual information. Contextual information is information gathered from a continuous assessment of the current environment. Aspects of the library environment, such as technologies and user experiences, are replaced each year with new innovations and new expectations. Information gathered in the past can help guide decision makers, but library professionals must also keep current with vendors' exigencies, their colleagues' priorities, and users' expectations. Their choices impact our decisions and our decisions impact their future choices.[7] When library decision makers rely solely on previous experiences or information gathered in the past, they ignore the contextual information present in the current environment, which unnecessarily restricts the effectiveness of the services their decisions create.

Patrons are one of the best examples of how a pool of information important to decisions can only be understood in context. Library professionals should not only know what users want or need but should know how patrons perceive the library within their own lives. Information in context contributes to whether the professionals can create services that keep the library relevant to our users. Accordingly, decision makers must be mindful not only of tracking the usage of electronic products and assessing the experiences of public service staff who work directly with patrons, they must also be acutely aware of the importance of learning from patrons themselves about their expectations.

Data tracking allows us to get closer to users' research processes through usage data, Web logs, and other statistical sources, but these are indirect and show only numbers, not causes and complex relationships between resources and users' needs. Indeed, these statistical data are not helpful in determining, for example, why use is less or more, whether the number of downloads actually contributed to the quality of research, or whether some service is missing altogether. Usability tests, library advisory groups, focus groups, direct observation, or other, more informal, means of understanding user experience are also critical for gathering appropriate information about user needs.

Contextual information gathering looks beyond the process that already exists and creates a more relevant decision based on new information from the environment. Adding new information to linear processes alters established procedure, which only makes decisions complicated and reduces efficiency. In dynamic processes, all identified stakeholders represent existing pools of information and effective outcomes must include these experiences as part of the decision-making process. As it does not need to be codified in a separate procedure, the information remains stored in the ongoing experiences of the stakeholders. In practice, procedures and decisions do need to be communicated to staff. Many solutions today including the myriad of social and collaborative software allow for written communication that can be easily modified by all involved when needed, allowing for dynamic processes and contributions from stakeholders. Decision makers can then rely on the collaborative process to make decisions with current information from the environment.

Collaboration: Informed Decision Making

Collaboration with stakeholders including library staff members is a critical component of acquiring contextual information and decision making. Organizations make decisions in a variety of ways based on the situation and the organization's philosophy. It is important to briefly look at several models of decision making to determine where an organization is and where it should move. Using a collaborative approach to making decisions results in more people who are behind the decision and willing to work to make that decision succeed.

Management writings generally divide decision making into four main patterns: command, consult, vote, and consensus. The first three decision patterns listed are all to one degree or another considered closed systems of thinking preferred by hierarchical organizations. A closed system works within linear, defined processes. Decisions are made with the goal of adhering to the priorities of the existing system, exempting procedures from most challenges in order to maintain consistency, and using familiar information already within the system to tweak, but not overhaul, the preexisting processes.

Above all, command decisions are efficient. These decisions are made by one person or a limited few, and therefore allow for quick decisions un-muddied by many conflicting experiences. Owing to the narrow source of information, this type of decision making is perilous for the decision maker, who must accept all repercussions for a failure. Once a command decision is made, effort must go into enforcing compliance with the decision among all those whose experiences were not included, or only included through the filter of the decision maker. Gaining "buy-in" is a popular way to refer to this necessity.

Consulting decisions are still efficient but, in contrast to command decisions, consulting decisions come with a semblance of including others in the decision-making process. In consultation, the decision maker asks for the input of others, but the decision will still be made by one person based on his or her own interpretation of the information available. This caveat still divests others of a sense of true influence, and, when this input is ignored, those asked go into the next round of consulting decisions a bit jaded as to the value of their input.

Voting, a third decision-making pattern, still puts the group under the expediency of accepting authority decision, this time of the majority, yet it comes closest to directly including more than just a primary decision maker in a decision. Voting, however, creates a group whose input is not reflected in the final outcome of a decision. Indeed, the either/or nature of voting ignores the knowledge contained in the losing decision, and accordingly, fails to take complete advantage of pooled knowledge. More often, politicization and faction-forming are very likely to occur and thus negatively influence the work environment and validity of future decision making.

As was indicated in previous text, these three patterns of decision making are all closed systems. They are closed in the sense that minority experiences do not stand on their own as influencing factors in the decision. A central experience, both in the form of an authority figure or of group majority, governs what is important to the decision and may even see the minority experience as threatening and destabilizing. In the dynamic reality of e-resource management, closed systems for decision making are inadequate. Indeed, here the accepted and familiar has a high likelihood of being "wrong" in the sense of not reflecting current environmental conditions.

Consensus through collaboration is the recommended decision-making process for dynamic processes such as e-resources management. A decision made through consensus may take more initial effort, but once made, it is more likely to be fully implemented since all parties involved helped to shape it. The authors would like to promote in consensus the sense of collaboration rather than simply compromise. Consensus includes a breadth of stakeholders' opinions, yet it prefers dialogue over compromise in its search for collaborative decisions.

Dialogue is the search for the third solution, unknown to either party at the start, yet created by both as they share their experiences and create a solution that truly reflects the reality of the environment. Dialogue first requires understanding the needs and experiences of each person involved. The richness of this pooled information moves the group into informed decisions that produce valuable outcomes for patrons.

Collaboration requires communication skills that can be new to command environments: group facilitation, personal awareness (because one now has personal value in organizational decisions), and interpersonal communication, all of which are beyond the scope of this chapter. However, collaboration also lays the groundwork for the most effective way to resolve these difficulties within an organization, adaptive learning through collaborative groups.

Successful Collaborative Teams

This section examines the characteristics of successful collaborative teams that harness the dynamic interchange inherent to adaptive learning and create the best outcome for libraries' services to patrons. E-resource management requires an environment of adaptive learners who pool information and make decisions through consensus. To better illustrate the concept of a collaborative team, see Table 5.2, which compares characteristics of collaborative groups and traditional committees.[8]

Collaborative teams are highly interactive and rely heavily on other group members to accomplish goals. Traditional committees often have one leader who must take on a major portion of the responsibility.

TABLE 5.2. Traditional Committees versus Collaborative Teams

Point of Comparison	Traditional Committees	Collaborative Teams
Purpose	Consult—give input into decision making by authority	Consensus—shared governance, information flow
Leadership	Leader appointed based on rank or status	Leadership roles determined by group, relevance to issue at hand
Responsibility	Rests on leader	Rests on each individual member of the team
Authority	Limited—subject to recommendations or command decisions from above	Significant—allowed to execute decisions, learn from mistakes
Degree of interaction	Low level, formal	High level, informal
Interdependence/reliance between group members	Low	High
Rewards and recognition	Individually based	Team based
Degree of "turfdom"	High	Low

Source: Based on Brian Quinn, "Understanding the Difference between Committees and Teams," *Library Administration & Management* 9, no. 2 (Spring 1995): 111-116. Reprinted with permission.

Group members do not have as much responsibility and there is a low degree of interaction with other members. In this type of working situation, the members are less likely to be truly involved in the work being done.

Current literature finds that the most successful organizations are flexible and innovative. These organizations achieve success through the promotion of work groups that are

- short-lived;
- goal-oriented;
- ready to assemble and reassemble to meet the changing needs of the organization and customers;
- willing to experiment and learn from trial and error;
- willing to tap into the necessary people or information to get the job done.[9]

Collaborative teams reflect many of these characteristics and work-ing with this type of team will make successful decision making quicker and easier. Essentially, one is tapping into the pooled knowledge of a group to decide together rather than deciding separately in the com-fort of individual areas of knowledge and procedure. With all the in-formation and constant changes swirling around a library's products and services, only a group of motivated learners can stay on top of the dynamic decisions that keep the library relevant. Shared responsibility, participation, and involvement are essential to the success of innova-tive work that leads to effective services for patrons. The unintended benefit is that more staff will lend their hearts and minds to projects when they see the direct positive impact of their ideas, untrammeled by command decision yet informed by the pooled experiences of col-leagues and patrons.

Individuals and groups who are adaptive learners are working in real libraries creating collaborative environments for managing e-resources. Some organizations have created formal means of group collabora-tions while others have created small, quick working groups that come together when a problem arises and disbands when the problem is solved. Two examples of the use of collaboration or collaborative teams to facilitate e-resource management are set out in the following text.

Georgia Institute of Technology's Library and Information Center has spent the past few years focusing on organizational change in regard to managing e-resources. The result is that two electronic re-sources coordinator positions now exist. One of these positions is in Acquisitions Services, where this person leads the Electronic Resources Unit of that department, negotiates licenses, and works closely with the cataloging and collection development departments. The other electronic resources coordinator is located in the collection develop-ment department and focuses on usage, evaluation, trials and trainings, keeping up with the latest new products and building relationships with e-resources vendors and the Digital Initiatives department. What is unique about these jobs is that cross-department collaboration and communication is part of the job descriptions. Both librarians are to keep each other informed, bringing together different perspectives to create new, innovative solutions to problems.

Some libraries do not have the budget, staff, or organizational struc-ture required to create new positions meant to work collaboratively, but they can benefit from the solution presented by Julia B. Dickinson and Sarah E. George at the 2004 NASIG Conference. These librarians saw that the existing formal structure of Ames Library at Illinois Wesleyan University was not an effective way to handle the constant change and new decision making involved in their library's digital environment. The library created working "hot groups" that came together to work on specific topics, gathered information (through literature reviews and discussions), and implemented change as needed. The groups are described as "highly informed and motivated." Key features of "hot groups" are that they are organically grown, task-driven, and short-lived with members who are motivated, interested, creative, and willing to take risks.[10]

Promoting Collaboration

These examples represent just a few methods of collaboration within an organization. However, both these situations reflect how an organization can focus on reorganizing to benefit library personnel serving the institution. In fact, the most critical investment in making a transition to an open, collaborative, adaptive learning culture is an organization's people.

An organization invests in collaboration when it nurtures the pooling of information within flexible work groups, rewards staff for a willingness to risk the mistakes of learning, and promotes the primacy of working with others for the purpose of improving the organization and its services. Within this learning culture, everyone from the director to the front lines takes initiative because the organization expects and supports it. In short, an adaptive learning culture occurs in an organization where everyone is a leader empowered to improve services, increase creativity, and foster innovation.

Table 5.3 compares the skills of a leader with the skills of an authority figure in an organization. Authority figures' actions are described using words like "controls," "executes," and "commands" while leaders "inspire," "initiate," and "explain." A learning organization has leaders at every level.[11]

From the library director to the front lines, from the supervisor to the supervised, we can all adopt the skills of a leader. Just as the linear process has its place as a stabilizing factor for handling known entities,

TABLE 5.3. Comparison on Authority Figures and Leaders

Authority Figures	Leaders
Control	Inspire
Perpetuate status quo	Challenge status quo
Short-term perspective	Long-term perspective
Execute	Initiate
Focus on doing things right	Focus on doing the right things
Command	Explain
Defend contractual relationships	Develop personal relationships
Punish deviance	Reward innovation
Stick to processes already in place	Question ineffective processes and policies
Value results over people	Value people as the source of results
Push information and decisions to others	Pull information from others for decisions
Decide with Command, Consult, and Vote	Use Consensus to make decisions

Source: Based on David Carson, Kerry et al. 1997. *The ABCs of Collaborative Change: The manager's guide to library renewal.* Chicago: ALA, p. 45. Reprinted with permission.

the command decision has its place as efficient and quick in action. Yet the decisions individual leaders make within their spheres of influence ultimately affect the quality of patrons' experiences. This relationship between making decisions and meeting users' needs places every member of the organization in a position of leadership. The organization in a dynamic technological environment can take advantage of this influence by nurturing the leader in each employee, and individuals can rest assured in their perceived importance within the decision-making structure of the organization, all with the effect of harnessing the strength of collaboration to effective decision making and library service excellence.

Change in the Individual

An entire organization does not have to change in order to make individuals introduce a more collaborative process into their own workflow. An individual does not have to be a manager in order to call together a few people to discuss a decision, walk down the hall to gather information from a different department, and suggest a small-scale test of patron responses before making a decision. An entire organization may not be interested in creating an environment of change in all areas because the risk seems too great or the change too threatening. Indeed, when an individual begins to collaboratively innovate, some colleagues may feel their established routines threatened. While collaboration takes a certain amount of courage and perseverance, the reward is the consistent joy of meeting the patrons at their point of need with outstanding service. This consistent success has a way of gradually influencing larger and larger numbers of colleagues within an organization.

Individuals cannot make others change, but they can become adaptive learners themselves, and their successes can become a model for others. The strongest point of leverage is one's impact on another's work. If we learn what would make our colleagues' work easier and consciously incorporate a positive effect on their workflows into our overall decision design, we will be taking a small step toward the positive influence adaptive learning can have on managing the dynamic change that is so challenging in e-resource management.

CONCLUSION

Managing electronic information will continue to grow in its influence on library processes. Adaptive, cross-functional teams that gather information, welcome the input and participation of stakeholders, share knowledge widely, and willingly explore the learning necessary for novel solutions will have a ready structure for managing the new, non-linear processes that e-resources introduce. The largest benefit rests in the fact that the library remains relevant to the patrons it serves. Within the flexible structure of collaborative teams and an adaptive learning mindset, library professionals will come to see changes not as stressors but as opportunities to improve their own work processes and their organization's service to patrons.

NOTES

1. Timothy E. Jewell et al., *Electronic Resource Management: Report of the DLF ERM Initiative* (Washington, DC: Digital Library Federation, 2004), http://www.diglib.org/pubs/dlfermi0408/ (accessed April 26, 2006).

2. Peter M. Senge, *The Fifth Discipline: The Art and Practice of the Learning Organization* (New York: Doubleday, 1990), 187-204.

3. Joan E. Conger, *Collaborative Electronic Resource Management: From Acquisitions to Assessment* (West Port, CT: Libraries Unlimited, 2004), 8-9, 19-48.

4. Peg C. Neuhauser et al., *Culture.com* (New York: John Wiley & Sons, 2000), 283.

5. Kerry David Carson et al., *The ABCs of Collaborative Change: The Manager's Guide to Library Renewal* (Chicago: ALA, 1997), 64.

6. Senge, *The Fifth Discipline*, 245-246.

7. Conger, *Collaborative Electronic*, 21-24.

8. Brian Quinn, "Understanding the Difference between Committees and Teams," *Library Administration & Management* 9, no. 2 (Spring 1995): 111-116.

9. Neuhauser, *Culture.com*, xv-xviii.

10. Lisa S. Blackwell, "Using Collaboration to Counteract Inertia in the Small Library," *The Serials Librarian* 48, no.3/4 (2005): 335-338.

11. Carson, *The ABC's*, 45.

Chapter 6

Staffing Trends and Issues in E-Resource Management

Maria D. D. Collins

INTRODUCTION

Much of the literature surrounding Electronic Resource Management (ERM) focuses on issues related to tools, processes, and workflow. Of course, the personnel that use ERM tools, design processes, and carry out workflows also form an important consideration. This chapter explores how personnel handling ERM tasks have fared as libraries make the transition to acquiring and managing e-resources. A two-pronged approach has been taken to examine trends and issues related to staffing and ERM including an informal survey used to provide indicators of these trends and a brief literature review to expand on these concerns. Even though the respondents of the survey do not represent a significant percentage of academic libraries, the survey's findings provide an effective springboard into a discussion of these issues.

First, however, the stage must be set to understand the context of the issues and trends being explored through a discussion of factors driving change in staffing for ERM. The next section of the chapter briefly focuses on the methodology of the staffing survey, explaining the logistics and the research questions the survey hoped to answer. To conclude, the survey results are presented in combination with the literature available on the topic in order to outline evident trends such as the blurring role of the paraprofessional, created and existing positions responsible for ERM tasks, the need for collaboration, and options for reorganization.

FACTORS DRIVING CHANGE IN PERSONNEL

There are many factors driving personnel trends in libraries. One of the most fundamental factors affecting staffing for ERM is the proliferation of e-publications. Through aggregated collections, publisher packages, and various other sources, libraries have become inundated with a critical mass of e-journals that requires increased personnel for effective management. Curtis further emphasizes the impact of the volume of e-journals on staffing, stating that

> to provide a significant quantity of electronic journals to your users means that even if you can outsource some of the management tasks, a large proportion of your library's staff will still have to take on some new responsibilities, and some positions may need to be completely re-configured.[1]

Note that Curtis also mentions user demand in the context of this quote. User expectation of immediate access is an additional factor influencing libraries' decisions to make personnel changes. Patrons today are heavily influenced by the Google world, where a search engine can help meet their information needs through natural language searching instead of restricted language or specialized searching such as Boolean logic. The ease of searching for materials in combination with the ability to immediately access content on the Internet increases patron expectations of seamless access while searching a library's resources. Highby explores this idea during an interview with Jill Grogg, electronic resources librarian at the University of Alabama. Grogg "agrees that user expectations for seamless access 'put tremendous pressure on technical services staff.'"[2] Of course, those of us creating the behind-the-scenes access solutions for our libraries recognize the truth of this statement, especially those of us that troubleshoot access problems for patrons. Many library staff may even go so far as to say patron demands are unrealistic given the level of personnel their library has devoted to ERM tasks.

One major reason why libraries have not devoted more staff to meeting users' demands for easy access to e-resources is libraries' adherence to traditional workflows and staffing responsibilities. For example, in Cherly Martin's case study examining how McMaster University Library reorganized to provide better access to e-resources, she discusses how traditional practices prevented the cataloging department from meeting patron demands and quickly providing access to e-resources. She states that "these problems were not anyone's fault, merely the end product of many years of work carried out without any critical examination of the underlying processes and standards."[3] This example illustrates one of the many instances in which traditional workflow practices have been ineffective in allowing staff to quickly handle the complexities of e-resources. The nonlinear nature of the e-resource life cycle requires unfettered procedures and practices for effective management. This means that libraries have to evaluate every process, workflow, or guideline in place and remove unnecessary procedures that no longer have value for patrons. Doing so gives the library the flexibility to create new workflow processes and free staff from nonessential duties that hamper their ability to prioritize e-resource tasks. In respect to the example provided in previous text, in December 2006, McMaster University Library's evaluation of workflow processes lead to the library "getting out of the cataloging business altogether." This announcement was noted by McMaster's university librarian, Jeffrey Trzeciak, on the library's blog. After reviewing cataloging workflows, decisions were made to outsource cataloging functions, redesign existing positions, transfer other positions to other departments, and introduce new, more relevant positions to the library.[4] This situation may be extreme, but it does demonstrate the workflow evaluation process as well as the creative staffing solutions that may be necessary to meet a given library's e-resource needs. To better utilize their human resources, libraries have to think critically about traditional workflows and thereby ensure that they are not trying to fit the square peg of e-resources into the round hole of existing workflows.

One final factor contributing to the need for personnel changes is the complex nature of e-resource work. In her 2002 study of staffing for e-resources, Duranceau aptly summarizes one response, stating that

> the demands that electronic resources place on staff are qualitatively different than the demands of print, in that "technology

has raised expectations and has also added a layer of complexity in the delivery of information that requires greater expertise among staff."[5]

Realizing this statement to be true, many libraries have tried to create a variety of stopgap measures (creating specialized positions or adopting ERM tools) to handle e-resource processes while still maintaining the status quo for processing traditional library materials.

In summary, libraries currently exist in an environment in which growing user demand has resulted in a proliferation of e-resources. The management of these resources has introduced new complexities, which traditional workflows cannot accommodate. To address these challenges, various staffing trends have arisen over the past decade. The staffing survey described in following text and explorations of the literature reveal several strategies utilized by libraries to combat the existing pressures for effective ERM in today's library environment.

SURVEY PURPOSE AND METHODOLOGY

In order to investigate potential trends and issues impacting personnel assigned to ERM responsibilities in academic libraries, a survey was sent out in August 2005 to the library directors of forty-two academic libraries having membership in the Association of Southeastern Libraries (ASERL). The primary research questions investigated by the survey are summarized as follows:

- What positions are handling e-resource work?
- Which departments are handling this work?
- How many of these positions are created and how many are existing?
- What tasks are each of these positions responsible for?

Out of the forty-two libraries contacted by e-mail, thirteen responded; eleven of those responses outlined specific positions that are primarily responsible for e-resource tasks at their institutions. Data from the responses have been analyzed and charted and will be discussed according to individual trends/issues in the next section. An example of the survey instrument is included as Appendix 6.A. at the end of this chapter.

TRENDS AND ISSUES IN STAFFING

Blurring Role of the Paraprofessional

One significant trend apparent from the survey and literature review is the blurring role of the paraprofessional or nonlibrary degree professional. Paraprofessionals or support staff have become responsible for many functions in the library that were traditionally a librarian's domain. Within the literature, Lopatin discusses a variety of case studies and reports that demonstrate increased staff involvement during the 1990s in a variety of departments including reference or information desk, copy cataloging, original cataloging, and collection development.[6] Two factors influencing this trend include lack of funding and increased patron expectations.[7] In addition, she specifically indicates a general trend in academic libraries toward the "empowerment of support staff."[8] The evolving role of the paraprofessional has both resulted in and been necessitated by the growing managerial responsibilities for academic librarians. El-Sherbini states that "professional librarians [have found] themselves managing departments or smaller units within departments, or coordinating activities in their departments."[9] In addition, librarians are focusing more on professional responsibilities and leadership roles in the library.[10] Given these circumstances where support staff in traditional areas of library management are taking on more complex responsibilities and librarians are taking on managerial roles in and outside the department, how have these trends influenced staffing for e-resources?

Initially, it appears that libraries were reticent to follow this same pattern of staffing for the management of e-resources. In the 1990s, e-journals and e-resources represented a much smaller portion of the collection than traditional library materials such as print. E-resources that did come through for processing were quite complex and did not fit within the normal range of workflow procedures. Often, one or two people, usually librarians, became the experts for handling these exceptions, thus creating a precedent in which knowledge of ERM issues became isolated and the librarians became information silos. The concentration of these e-resources tasks in one or two librarians resulted in the creation of the electronic resources librarian. In fact, in a study of job descriptions appearing in *American Libraries,* Fisher noted that shortly after e-resources were introduced to libraries in the 1990s, the first electronic resources librarian positions appeared in 1992. Between 1992 and 2001, a total of 298 job postings appeared in this same publication for electronic resource librarians.[11] This is a substantial increase in postings for one type of position in a short period of time.

Even within the last few years, when many libraries have experienced a tremendous increase in their e-collections, many libraries have failed to distribute e-resource tasks. Jasper and Sheble provide an example of this scenario at Wayne State University. They state that, "prior to 2003—and despite momentous changes in the scale and variety of its e-resource offerings—the Library System had no individual staff member whose position was principally (much less solely) devoted to e-resources management."[12] Of course, in defense of Wayne State, any manager of e-resources will be the first to say that distributing e-resource responsibilities is easier said than done. Issues of training, lack of routine tasks, and the volume of work to complete in short periods of time often prevent librarians from integrating e-resource processing throughout library units. Fischer and Barton effectively characterize this situation in their description of libraries' first interactions with e-resources. They state that

> When libraries first began to deal with electronic resources, all of the work was done at the librarian level because of its novelty and complexity. Now it seems apparent that the work needs to be distributed out to high-level support staff, but librarians seem to be struggling to identify what tasks are best to delegate.[13]

Part of the problem with delegating e-resource tasks can be linked to the volatility and nonlinear nature of these materials. It has taken some time for librarians to analyze the exact steps necessary for a successful e-resources workflow. However, with this knowledge comes an understanding of how to allocate staff resources.

Obtaining an understanding of who is doing what in respect to e-resource tasks is one of the objectives of the personnel survey discussed earlier. Even with a small number of libraries responding, it is interesting to compare the number of librarians with the number of paraprofessional positions that are primarily responsible for e-resource tasks. The survey specifically targeted responsibilities in the following areas: new subscriptions/deals and trials, licensing, registration and authentication, cataloging, e-journal management tools, acquisitions, and usage statistics. Demographics for each position were also requested, including the minimum education required for the position. For the purposes of this discussion, positions requiring a graduate degree (Master of Library Science (MLS) or other) will be grouped as librarians, and positions requiring some college or an undergraduate degree will be grouped as paraprofessionals. As is apparent in Figure 6.1, 75 percent of the fifty-six positions described as being primarily responsible for e-resources were librarians and 25 percent were paraprofessional. These results indicate that even if support staff are becoming more of a presence in respect to staffing for e-resources, librarians are still handling the more complex tasks.

Although a follow-up survey would be required to determine with certainty if the number of support staff with ERM responsibilities will continue to grow, a review of the literature suggests that this delegation of responsibilities is occurring. For instance, in respect to the Wayne State example cited in previous text, Jasper and Sheble note their library's intention to increase the number of paraprofessionals handling e-resource tasks. They state that "in the coming year our goal is to examine all technical services workflow with an eye toward shifting

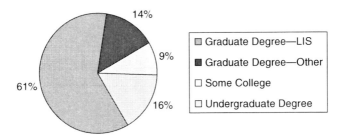

FIGURE 6.1. Education Level for E-Resource Positions

the focus of our work so that—as with the materials budget—the majority of staff time is spent dealing with electronic resources."[14] This may very well be an accurate prediction of many academic libraries' intentions as they continue to plan for the growth of their e-resource collections. Having isolated experts manage and process what for many libraries represent over 50 percent of their materials budget is an unsustainable model. Expanding the role to the paraprofessional to take on more ERM tasks is a potential solution.

Created versus Existing Positions

Another trend evident in today's academic library environment is the creation of new positions primarily responsible for e-resource tasks. Browsing through the latest issues of the *Chronicle* will most likely reveal a handful of positions with a variety of e-resource responsibilities in their job descriptions. In fact, in a recent *Library Technology Report* discussing staffing for ERM, Geller discusses a staffing survey from the Association of Research Libraries' *ARL Spec Kit* 282. She notes that most of the libraries responding to this survey have made some kind of staffing change in order to manage e-resource tasks and "overwhelmingly, the responding libraries had created new positions."[15] There are a host of reasons that necessitate the hiring for these positions including the volume of e-journals being acquired by academic libraries, the complexity of the workflows involved and the time-consuming nature of the work. Gardner notes that "libraries everywhere are finding that e-journals involve more staff and increased staff time at both the acquisitions and maintenance stages of the work flow process than their printed counterparts."[16]

The personnel survey conducted for the purposes of this chapter also indicates that many libraries are creating positions to handle e-resources. Of the fifty-eight positions described in the survey, twenty-three were created. Of these positions, five were transferred from another department in the library: one from reference to systems, two from reference to collection management, one from cataloging to systems, and one from cataloging to scholarly communication and integrated digital services. It is interesting to note that the majority of these positions were created within five years of when the survey was conducted (i.e., between 2001 and 2005). See Figure 6.2. These

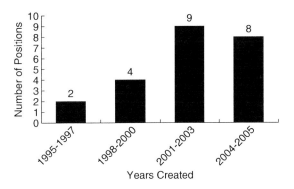

FIGURE 6.2. Recent Growth in E-Resource Positions

numbers reflect an increasing demand for positions with e-resource management skill sets. They also reflect a shift in organizational priorities for some of these libraries to realign staffing resources to better handle the varying demands e-resources create.

In addition, numerous departments benefited from these positions; there was no primary concentration of new positions in any one area. New positions were added to the following departments: cataloging, acquisitions, serials, electronic resources, reference, systems, scholarly communication, and collection management (See Figure 6.3). The variety of departments represented reflects the wide scope and nature of e-resource work and the distribution of these responsibilities across libraries. These data also indicate that there is no set pattern of departments that libraries select for ERM. For the eleven libraries that submitted position descriptions, no set type of position is being created across libraries; instead, these new positions are being designed to handle a wide variety of tasks.

While examining the personnel survey's numbers for created positions, it becomes obvious, however, that the majority of the positions described are existing. Thirty-five of the fifty-eight positions described are existing positions that have evolved to take on ERM tasks. The literature heavily supports the notion that, even as libraries are adding positions, they are more often using existing positions to carry out new responsibilities. For example, Gardner notes that "one of the trends in staffing as a result of the increased workload imposed by e-journals is

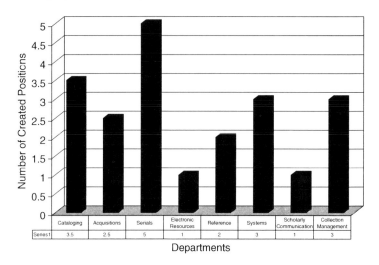

Series1	Cataloging	Acquisitions	Serials	Electronic Resources	Reference	Systems	Scholarly Communication	Collection Management
	3.5	2.5	5	1	2	3	1	3

Departments

FIGURE 6.3. Departments Benefiting from Created Positions

to incorporate new responsibilities into existing library positions."[17] Another survey, conducted by Duranceau, also indicates the prevalence of existing positions handling e-resources tasks. She states that

> most of the libraries responding to our survey have, as noted above, added significantly to staff working on electronic resources, but most have done so by distributing the work among many additional players rather than hiring staff to handle the particular demands of digital resources.[18]

Therefore, considering both the indications from the personnel survey conducted for this chapter and the discussion of the literature noted in previous text, it appears that libraries are considering a variety of staffing solutions to address the changing workload; libraries are reinvesting in existing staff to evolve job responsibilities and adding new staff to fill in gaps for other e-resource needs.

Unfortunately, even with both created and existing positions supporting ERM functions, staffing is still insufficient to handle the growth of e-resource collections that most libraries are experiencing. This was one of the primary findings of Duranceau's informal investigation of staffing for ERM. She notes that "libraries have clearly made fairly significant efforts to reallocate staff or redefine positions, but if this group of libraries is at all representative, these efforts have not been adequate to meet the rather astonishing level of demand created by the volume and complexity of digital collections."[19] In other words, many libraries are being forced to do more with less. To address this dilemma, existing library functions will have to be analyzed and prioritized to ensure that libraries are directing their staff resources toward the materials having the greatest demand by patrons.

THE NEED FOR COLLABORATION

In addition to handling more responsibilities with less staff, libraries often distribute ERM responsibilities over multiple departments. A case in point: eleven different departments were affiliated with the job positions across all libraries described in the personnel survey (see Figure 6.4). At the least, effective e-resource management requires input from multiple departments. The participation of multiple departments per library is also reflected in the personnel survey results. Out of the eleven libraries who described positions, all but one included

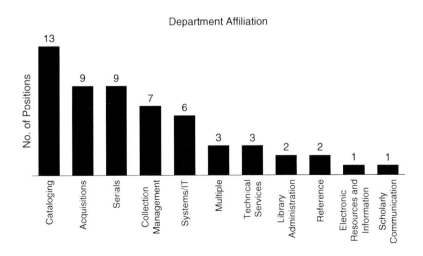

FIGURE 6.4. Departments Handling ERM Tasks

position descriptions from more than one department. Five included positions affiliated with three different departments, one noted a range of positions affiliated with four departments, and two libraries discussed positions affiliated with seven different departments. This kind of infrastructure (whatever the organizational model) requires effective communication channels to ensure consistency and minimize duplication.

There are several communication strategies discussed in the literature that facilitate cross-department communication. Geller discusses a team approach that can be used alongside a distributed model for managing e-resources. In this model, responsibilities are integrated into existing departments already handling particular functions such as selection, acquisitions, or cataloging. An organized team representing members of the departments that make up the distributed management chain serves as a coordination and communication arm that works alongside the given organizational structure. Geller notes that, "without this cross-departmental team strategy, the distributed approach (to managing e-resources) is in danger of becoming fragmented and existing without a clear communication mechanism."[20] There are additional advantages to this team structure outlined by Curtis. She states that a complimentary communication framework "will allow your library to experiment with different configurations and affiliations without your having to redo personnel documents or disrupt reporting structures."[21]

Another strategy to enhance communication across departments for e-resource management is the use of collaborative technologies, such as a Wiki, blog, Google spreadsheet, or a dedicated listserv. The use of these emerging technologies allows individuals to keep up with initiatives and communicate quickly. For example, Jasper and Sheble point out advantages to using a listserv, noting that this method provides "a forum for discussing issues as they arise and [can serve] as a mechanism by which staff working in different functional areas can share expertise and learn from each other's experiences."[22] Whatever method a library selects, it is clear that communication must be a priority. Individuals working with e-resource material need to work across departments and with vendors, publishers, and other libraries to ensure that these materials are managed effectively. Establishing communication mechanisms that are incorporated into staff routines can only enhance this collaboration.

OPTIONS FOR REORGANIZATION

Another important issue with respect to staffing is the evolving organizational structures libraries are utilizing to handle e-resources. This topic was not addressed specifically in the personnel survey, but as indicated in previous text, the survey does show that numerous departments per library are involved in the management of e-resources. One survey respondent did not provide position-by-position descriptions but noted that her library was in the process of reorganizing to distribute duties across departments, and that no particular positions were dedicated to e-resources at the exclusion of other positions. Discussions of similar reorganizations are prevalent in the literature with several models predominating.

The first model, the e-resources department model, has served as a stopgap measure for many libraries. The specialized knowledge needed for e-resource management and the quick turnaround time required for processing these materials makes the idea of a separate e-resources department that can handle most aspects of the e-resource life cycle very attractive. Furthermore, print serials staff with related expertise may have ongoing responsibilities to manage the print collection that still require a substantial portion of their time if a library has not yet switched a majority of its subscriptions to online only. Therefore, this model works well in libraries with small e-collections, but as these collections grow, this model becomes difficult to sustain. Curtis aptly describes the scenario that many e-resource departments may find themselves in as they transition to e-only collections:

> Managing e-resources requires a range of expertise, so you will have to reassign staff from acquisitions, cataloging, serials, collection development, reference, Web-development and systems areas; and, as electronic resources and services continue to expand, this new department might ultimately devour a large portion of the library.[23]

This statement is far from the exaggeration it first appears to be, indicating that the viability of the e-resources department model may be short-lived depending on a library's collection goals. However, this type of organizational model may be appealing for libraries that perceive e-resource responsibilities as "add-on" functions to existing workflows.[24] Essentially, the "add-on" concept of segregating e-resource responsibilities and expertise for ease of management is counterintuitive to the concepts of an integrated model, which would incorporate e-resource workflows into existing processes. Curtis's statement in previous text implies that an integrated approach is often best suited for management of large electronic collections. Libraries must achieve a careful balance between the most efficient model needed to manage their existing collections and a future model that may better facilitate the demands of growing electronic collections.

This brings us to the next model, the integrated or distributed model discussed in the previous section. In this model, e-resources are not treated separately but are instead integrated into a library's existing organizational structure. Considering that established workflows for traditional materials such as print are not adequate for handling e-resources, this is not always an easy adjustment. Careful management of the integration process in addition to an extensive training plan is needed to accommodate these new processes and ensure that existing staff are competent. The model does take advantage of established expertise specific to library functions like acquisitions and cataloging. The experts in these areas simply have to be willing to expand their role by applying their expertise to resources in a new format. Geller notes that this model can be particularly effective for libraries that are cancelling print subscriptions in favor of online access, stating that the "lower level of activity in print-subscription management is freeing up staff time for new responsibilities."[25] In the long run, this may well be the most realistic model. Indeed, in the face of the current or future critical mass of online subscriptions and the potential instability of new organizational structures that have not stood the test of time, the integrated model enables a library to redefine staff responsibilities while maintaining its existing organizational structure.[26]

Geller mentions a third model that is essentially a hybrid of the first two models described. This model provides managerial structure in the form of an e-resources librarian or unit, but the e-resource responsibilities are still distributed across the library.[27] This approach seeks to provide the best of both worlds by establishing managerial expertise but utilizing the established workforce per library function. Geller emphasizes that "this model takes into account both the new skills and staffing levels of the discrete electronic-resource management model and the unity concept of the entire library collection observed in the integrated model."[28]

One last model to note, incremental reorganization, is actually a method for assisting organizational change. This method is essentially a very deliberate version of the integrated model where e-resource responsibilities are carefully examined and slowly integrated into existing positions through attrition. According to Curtis, this model "can have a dramatic effect over time without the risk of traumatizing an organization or taking attention away from the challenges at hand."[29] This plan for evolving e-resource workflows is very controlled and methodical, taking advantage of vacancies and hiring opportunities. Job responsibilities are carefully evaluated and job descriptions are adjusted to accommodate new priorities. Curtis also mentions cross-functional teams to facilitate communication and the need to reward staff who are proactive in taking on new roles in the department.[30]

CONCLUSION

Libraries have reached a critical junction in respect to staffing for ERM. Many libraries have spent the last decade reacting to e-resources and being driven by factors such as e-collection growth, complex workflows, and patron demand. These reactions have resulted in valuable contributions to the libraries such as additional staff resources and new departments who have established e-resource routines. Many libraries now have a firm understanding of e-resource responsibilities and can take a step back to proactively plan for ERM. This chapter highlighted several staffing issues and concerns that should be considered during this planning process such as the utilization of the paraprofessional, the use of existing positions, and the pros and cons of organizational models. Ultimately, libraries will have to examine their plans for growing their e-collections, examine their current staffing resources and seek funding opportunities before determining the most appropriate plan for organizing their staff and library functions.

NOTES

1. Donnelyn Curtis with Virginia M. Scheschy, "Shifting Library Resources" in *E-Journals: A How-To-Do-It Manual for Building, Managing, and Supporting Electronic Journal Collections,* no. 134 (New York: Neal-Schuman Publishers, Inc., 2005), 91.

2. Wendy Highby, "Linking Changes Thinking: The Acceleration of E-Journal Control Issues for Technical Services Staff," *Colorado Libraries* 31, no. 3 (Fall 2005): 41.

3. Cherly Martin, "Workflow Analysis as a Basis for Organizational Redesign at McMaster University Library," in *Innovative Redesign and Reorganization of Library Technical Services: Paths for the Future and Case Studies,* ed. Bradford Lee Eden (Westport, CT: Libraries Unlimited, 2004), 202.

4. Jeffrey Trzeciak, "Getting out of the Cataloging Business," McMaster University Library: Partners in Teaching, Learning and Research, http://ulatmac .word press.com/2006/12/03/getting-out-of-the-cataloging-business/ (accessed January 26, 2007).

5. Ellen Finnie Duranceau and Cindy Hepfer, "Staffing for Electronic Resource Management: The Result of a Survey," *Serials Review* 28, no. 4 (2002): 317.

6. Laurie Lopatin, "Review of the Literature: Technical Services Redesign and Reorganization," in *Innovative Redesign and Reorganization of Library Technical Services: Paths for the Future and Case Studies,* ed. Bradford Lee Eden (Westport, CT: Libraries Unlimited, 2004), 8.

7. Ibid., 6; Magda El-Sherbini, "Technical Services Between Reality and Illusion: Reorganization in Technical Services at the Ohio State University Libraries—Questions and Assessment," in *Innovative Redesign and Reorganization of Library Technical Services: Paths for the Future and Case Studies,* ed. Bradford Lee Eden (Westport, CT: Libraries Unlimited, 2004), 384-385.

8. Lopatin, "Review of the Literature," 4.

9. El-Sherbini, "Technical Services," 395.

10. Lopatin, "Review of the Literature," 6.

11. William Fisher, "The Electronic Resources Librarian Position: A Public Services Phenomenon?" *Library Collections, Acquisitions, and Technical Service* 27, no. 1 (Spring 2003): 4.

12. Richard P. Jasper and Laura Sheble, "Evolutionary Approach to Managing E-Resources," *Serials Librarian* 47, no. 4 (2005): 669.

13. Karen S. Fischer and Hope Barton, "The Landscape of E-journal Management," *Journal of Electronic Resources in Medical Libraries* 2, no. 3 (2005): 63.

14. Jasper and Sheble, "Evolutionary Approach," 66.

15. Marilyn Geller, "ERM: Staffing, Services and Systems," *Library Technology Report* 42, no. 2 (March/April 2006): 12-13.

16. Susan Gardner, "The Impact of Electronic Journals on Library Staff at ARL Member Institutions: A Survey and a Critique of the Survey Methodology," *Serials Review,* 27, no. 3/4 (2001): 17.

17. Ibid., 26.

18. Duranceau and Hepfer, "Staffing for Electronic," 317.
19. Ibid., 320.
20. Geller, "ERM: Staffing," 13.
21. Curtis and Scheschy, "Shifting Library," 103.
22. Jasper and Sheble, "Evolutionary Approach," 66.
23. Curtis and Scheschy, "Shifting Library," 98.
24. Geller, "ERM: Staffing,"12.
25. Ibid., 12.
26. Curtis and Scheschy, "Shifting Library," 98-99.
27. Geller, "ERM: Staffing," 25.
28. Ibid., 23.
29. Curtis with Scheschy, "Shifting Library," 99.
30. Ibid., 99.

APPENDIX
SURVEY

STAFF/PERSONNEL RESPONSIBLE
FOR E-JOURNAL/E-RESOURCE TASKS

Demographic and Count Questions

1. University/Institution:
2. Library name:
3. Current FTE:
4. Number of current positions that are primarily responsible for e-journal/e-resource tasks (e.g., trials, licensing, authentication, cataloging, e-journal management tools, etc.).
5. Of the current positions that handle e-journal/e-resource responsibilities at your library, how many were created
   a. within the last ten years
   b. within the last five years
   c. within the last two years
   d. within the last year
6. Of the current positions, how many were existing positions that have evolved over the years to handle e-journal/e-resource responsibilities?

Position Description (repeat as necessary for each position primarily assigned to e-journal/e-resource responsibilities).

1. Created position or existing?
2. If position was created within the last ten years, what year was the position created?

3. Position title:
4. Department(s) affiliated with this position:
5. Title of person this position reports to:
6. Previous department affiliated with this position if not the same as question 4. Provide a brief description of responsibilities under previous department if different from current responsibilities:
7. Title of the person this position used to report to under previous department (if different from answer stated above in question 5):
8. Minimum level of education required for this position:
   a. High school
   b. Some college
   c. Undergraduate Degree
   d. Graduate Degree—Library or Information Science
   e. Graduate Degree—Other
9. Annual salary range for this position
   a. $10,000-20,000
   b. $20,001-30,000
   c. $30,001-40,000
   d. $40,001-50,000
   e. $50,001-Higher
10. In the list below, check all duties/responsibilities directly related to e-journal or e-resource management carried out by this position. (Check all that apply, add any duties not included under "Other.")

New Subscriptions/Deals and Trials

Investigate and recommend possible consortial deals to consider (e.g., ESIG, Solinet, statewide consortium, etc.)
Contact vendors and coordinate trials and demonstrations of the following:

E-journal packages and deals (i.e., Elsevier, JSTOR)
Databases (Abstracting and Indexing)

Notify appropriate library personnel of trial information
Collect feedback and make recommendation for purchase after trials or demonstrations
Suggest individual e-journal subscriptions to evaluate and add to collection
Evaluate e-journal subscriptions or deals (in case of no trial or demonstration)
Obtain pricing information from vendors
Negotiate pricing with vendors

Licensing

  Review license agreements
  Negotiate license agreements
  Sign license agreements
  Scan and/or file license agreements

Registration/Authentication

  Register e-journal/e-resource subscriptions
  Authenticate e-journal/e-resource access

Cataloging

  Perform copy cataloging for e-journals/e-resources
  Perform original cataloging for e-journals/e-resources
  Administer use of MARC record services (i.e., MARCit from Ex Libris,
    Full MARC Records from Serials Solutions)
  Maintain 856 fields, correct broken URLs

E-Journal Management Tools

  Research products/tools to select for trials
  Contact vendor and coordinate trials and demonstrations of product
  Collect feedback and make recommendation for purchase of product
    or tool
  Serve as link resolver administrator
  Perform data maintenance for link resolver
  Perform Electronic Resource Management (ERM) system maintenance
  Manage e-journal list/maintain electronic resources page
  Administer metasearch tool (i.e., Metalib, Webfeat)

Acquisitions Duties

  Create/place order
  Pay/process invoice
  Set up/maintain serial record (not cataloging record)
  Maintain/create holdings statements in ILS
  File paperwork related to e-journals
  E-journal claiming (check to see that electronic issues are available)

Usage Statistics

Collect login, search and request statistics
Review turn away statistics to determine number of simultaneous users
needed for library

Other

Note any additional duties not included in list above.

11. Provide a brief summary of current responsibilities not related to
e-journals.

Chapter 7

Partnering with the Patron

Beth Ashmore
Jaroslaw Szurek

INTRODUCTION

The advent of e-journals and databases provides users and librarians with an unprecedented opportunity to integrate the tools for finding information with the information itself. With these old resources in new formats, innovation has a new ally in the form of increased information about users' habits and preferences. The popularity of the Internet and a growing one-click mentality is driving librarians and vendors to collaborate in order to design products that provide more seamless access to resources and services. In addition, research on e-journals, databases, and users' information seeking behaviors has influenced both librarians' methods of description and organization and the development of library-specific products and tools. The increased amount of data about user behavior that is supplied by e-resources providers has enhanced the information professional's ability to design tools and organize information in a way that makes sense to the average user. This chapter will explore how the partnership between information professionals and information users (or library patrons) has shaped the integration and acceptance of e-resources and management tools throughout the last few decades. These interactions have provided invaluable insights into patrons' genuine usage patterns. Furthermore, users, whose roles have been historically reduced to passive information recipients, have benefited by becoming participants in the process of knowledge dissemination.

THE EVOLVING E-WORLD

How did we get here? The infiltration of e-resources into libraries is nothing if not paradoxical—completely subtle and insidious while at the same time altogether earthshaking. It is easy to see in retrospect how things have gotten to this point, but did it always seem so certain? With a glance back at librarian and other stakeholder perspectives of e-journals and online databases, ten to twenty years ago, it becomes obvious how great an impact these resources had on the way libraries provide access to information.

E-Journals

To read the early literature about e-journals is to run the gamut of human emotions. There was hope that this new format would alleviate the serials pricing crisis of the early 1990s. There was confusion over how to manage basic serials processing such as check-in, claiming, and cataloging. Since Web sites were brand new themselves, there was even bafflement about how users were actually going to discover and access these new resources.

Notwithstanding these considerable concerns the general consensus seemed to be that these resources would significantly change serial publishing. However, predictions about the speed or depth of their effect varied widely. Brett Butler, publisher of *Electronic Publishing Business* and, therefore, uniquely cognizant of the latest trends in this publishing area, presented in his 1986 *Serials Review* article "Scholarly Journals, Electronic Publishing and Library Networks: From 1986 to 2000," a vision of what the e-serials marketplace would look like a full ten years before things really began to change. Butler stated, "librarians' favorite media after print will continue to be microform . . . More collections of periodicals will be offered on high-resolution optical-disk storage systems after 1990, but these will not, in general, contain retrospective runs of serial holdings."[1]

Despite his surprising (by today's standards) belief in the unfading popularity of microforms, Butler was not wrong about the limitations of CD storage. For that matter, in a non-Internet future, the possibilities for e-journals could also seem limited. In addition, Butler is remarkably prescient about other innovations: "article-level indexing of journals is virtually comprehensive for most periodicals in the United

States . . . available in print and selectively online. This will be enhanced to public catalog access at the article level to the library's own
holdings."[2] Notably, Butler describes a primitive kind of linking in a
pre-Internet environment, but this is not all that surprising. The desire
to connect the information needed with the information available
through a user's local institution is a pretty basic and, therefore, welcome innovation.

It was not long after Butler's prophetic article that the library community was attempting to envision the future of electronic publishing, most specifically e-journal publishing. Two projects in the early
to mid-1990s demonstrated the hope that libraries saw in e-journals.
The Mr. Serials Process created by the North Carolina State University Libraries was a direct attempt to place libraries at the heart of the
e-journals industry in the hope that a more cost-effective model for
scholarly information acquisition would result. This pilot project collected existing information science-related electronic serials into an
FTP archive in order to demonstrate "an automated method of collecting, organizing, archiving, indexing and disseminating electronic
serials."[3] To hear the project described, sounds remarkably like some
of the Open Access projects around today: "the Mr. Serials Process
hoped to demonstrate to the scholarly community that if they publish
their findings in inexpensive or free peer-reviewed Internet-based
journals, then libraries can effectively help facilitate the scholarly
communications process."[4] The Mr. Serials Process is another example of librarians' attempts to use e-journals to respond directly to user
needs. In this case, the need was to have access to a broader range of
scholarly communication in an ever-growing journal market plagued
by dramatic price inflation.

Another notable e-journal project to come out of the early 1990s
that demonstrates a large and concerted effort to use this new format
to respond directly to user needs is the SuperJournal Project created
by the SuperJANET (Joint Academic Network) project on information
resources (SPIRS). This project represented an early feasibility study
of whether users would embrace e-journals and, if so, whether they
could become an alternative to print.[5] Taking advantage of the existing
high-speed network, in this case the United Kingdom's SuperJANET,
SuperJournal brought together the content of nine STM publishers
including academic societies as well as commercial publishers. This

project proved to be more successful in developing a viable model of network publishing than the more open-access-oriented Mr. Serials. Where Mr. Serials was an attempt to democratize scholarly communication, SuperJournal was a test of interfaces, bandwidth, and economics, using established print publishers and journal titles, unlike Mr. Serials' brand-new electronic titles. The success of e-resource hosting services such as Ingenta and Highwire is evidence that the SuperJournal model remains active in today's marketplace.

The big question that surrounded both the Mr. Serials Process and the SuperJournal Project was whether academics would want e-journals. Initially, the answer was not promising. Surveys of faculty by Budd and Connaway[6] and Gomes and Meadows[7] in 1995 and 1996 respectively indicated that faculty were less than stellar in their immediate endorsement of this new format. Faculty had doubts about the authority and stability of e-journals and were not willing to risk losing these attributes in their pursuit of promotion and tenure. Parallel publishing of print journals in electronic format would prove to be the driving force behind the legitimization of e-journals.[8] Parallel publishing provided the best of both worlds for users—all the prestige of the print journal combined with more efficient access. The STM community, publishers, and users, recognized the benefits of this model early and, whether it was licensing content to aggregated database services or creating homegrown databases of a publisher's titles, publishers were willing to enhance the market for their journals by providing this additional access to their products. For faculty, embracing e-journals did not mean an immediate disdain of all things print, but it did mark a change in the journal market; whether it would be the change librarians battling high journal prices wanted remained to be seen.

By 1997 the Internet had changed everything and commercial vendors had answered many of SuperJournal's feasibility questions with a resounding "maybe." In a 1997 Electronic Journal Market Overview for *Serials Review,* George Machovec acknowledged succinctly that, "the electronic journal (e-journal) is finally 'coming of age' as a result of the explosion of Internet use, particularly World Wide Web technology."[9] Machovec identified the two contenders in the emerging fight for the e-journal market as aggregators and publishers, demonstrating how quickly the open access concepts set forth by projects like Mr. Serials and even earlier by predictions by Butler, were overwhelmed by the commercial market. This was not to say that e-journals were problem free by 1997. In fact, Machovec also identified several major unresolved issues including pricing, layout, copyright, backfiles, reliability and access.[10] Librarians responded by becoming strongly committed to seeking solutions to these problems. As a result, a new undercurrent of research on e-resource usage patterns was building to a critical mass, helping librarians better understand how patrons' needs fit into the e-journal equation.

Usage studies were not new when e-journals came on the scene, but new attempts to study user behavior in the electronic environment were not your mother's usage studies. In the 1990s, researchers like Carol Tenopir, Donald W. King, and Carol Montgomery, Randy A. Hoffman, and Sarah E. Aerni began to develop theories on how researchers currently get information and how e-journals would change this process, if at all.[11,12] These researchers were providing the big picture of use patterns and, with the new availability of usage statistics, a brand new form of user feedback was born that did not require a single survey to be completed.

Previously, periodical usage statistics tended to be shoddy at best. Circulation statistics missed most browsing and photocopying patrons and, if journals were noncirculating, this method was not applicable. In *Management of Serials in Libraries,* Thomas E. Nisonger discusses the variety of techniques that have been developed over time to measure print periodical use from table counts to the slip method to direct observation. Nisonger also recognizes that these in-house methods come with some inherent difficulties, including the need "to define what constitutes use," the expense and time associated with implementation and the fact that "most studies depend on user cooperation."[13] Vendor-supplied usage statistics for e-resources, while imperfect in their own way, provide a new avenue for reviving this form of user assessment. The hard numbers provided by vendors give librarians an opportunity to reevaluate collections and, best of all, to enhance the older forms of user assessment that Nisonger describes. Librarians also have been given the ability to collect users' opinions in a less invasive way. By adding small question surveys or creating focus groups on these new resources, users have the opportunity to voice their opinions about the direction of the library collection.

It is easy to say that e-journals changed everything we knew about libraries. However, long before e-journals were embraced, the electronic environment was embracing another key resource, e-indexes, that would make e-journals seem inevitable to many users and librarians.

E-Indexes

When citation databases first entered the consciousness of researchers in the 1960s and 1970s, they represented everything that was good and bad about indexes and library resources in general. Early information services like Dialog and Orbit were powerful in the sense that they could help researchers complete previously time-consuming tasks in moments. They were exclusionary because users had to be one of the select few who had access to them and knew how to use them. During this period, librarians took on the role of searching databases for users; it seemed a logical step because librarians are experts at mediating information for users. This was a good move for the profession, once again making librarians indispensable, but it did not benefit interface design since librarians were already highly trained in adapting to varying organizations of information. Command line searching was just fine by them; however, database vendors were not encouraged to improve the interface for the common library user. In addition, academic and public library users were not their primary audience. As Butler pointed out in 1986, "even bibliographic services that are most closely tied to libraries—Dialog, Orbit, and BRS—conduct over 80 percent of their collective business with special or corporate libraries rather than research libraries."[14]

Once again it was not until the Internet, the great equalizer for information regarding all, that the information environment started to become more democratized. Librarians and database vendors alike realized what a boon it would be if everyone could search for themselves and suddenly, with the concept of Web-based article databases, the technology was there to do it. Tenopir et al., found in their May 2003 *D-Lib* article that during what the researchers have called the "evolving system phase" of access to e-journals from the late 1990s to 2003, "the patterns of article identification change[d]; . . . the proportion of articles found by online search [was] well up."[15] The researchers saw a shift in the way scientists were discovering the articles they needed—moving from less reliance on browsing journals to increased time spent searching online for articles. Particularly spurred on by the advent of aggregated databases, undergraduates were among the most enthusiastic audiences for databases. In a 2000 North American Serials Interest Group (NASIG) workshop, Jie Tian from California State University Fullerton demonstrated the popularity of full-text database searching among their populations: "of the 97 databases available, survey respondents used 74. Full-text retrieval was a definite focus; the survey showed that 82 percent of database usage resulted in the retrieval of articles. Survey respondents used aggregated, full-text, interdisciplinary databases . . . most often."[16]

The introduction and popularity of Web-based aggregated databases was decidedly a double-edged sword. In addition to furthering the construct that everything is available online, it muddied the waters between databases and Web search engines. From the beginning it appeared that databases would have a hard time competing with the World Wide Web in general. For example, a 2000 study conducted at Israeli universities revealed that approximately 60 percent of respondents reported using the Web every day while only a little over 10 percent of respondents used e-bibliography tools and e-journals every day. High percentages of the same users rated e-journals and e-bibliography tools as indispensable, but the Web remained a daily companion.[17] Oddly, the ongoing difficulty of communicating to users the difference between the Web at large and online library resources is made more difficult as libraries continue to innovate. Indeed, as libraries seek to improve the information-seeking experience by co-opting those qualities that users find so attractive in the Web, the differences between these library resources and general Web resources begin to blur. This blurring of the line between purchased library resources and the open Web is not necessarily a bad development except that it may reinforce some users' unfounded faith in the authority of all information that is found online. Some might suggest that it is important for library resources to look and feel different for this reason.

WHAT ARE OUR USERS TELLING US ABOUT E-RESOURCES?

When libraries were the only game in town, it was easy to ask users to learn the library's methods of organizing information. With the arrival of the Internet, however, users suddenly had a whole new set of options for finding information, which made the library's classification schemes, call numbers and print indexes appear outdated and foreign to users.

In 2002, the Pew Internet and American Life report, *The Internet Goes to College* told us that by ages sixteen to eighteen, 100 percent of the college students surveyed were using computers and ". . . the Internet was commonplace in the world in which they lived."[18] Not only had these students not necessarily been raised with more traditional library resources and skills, they did not necessarily see a need to acquire them. As the 2002 report points out: "nearly three-quarters (73 percent) of college students [said] they use the Internet more than the library, while only 9 percent said they use the library more than the Internet for information searching."[19] While using the Internet more than the library may seem like a red flag, it does not say that those students are not using the library, but simply that their preference is with the Internet. The real red flags come in the anecdotal observations found in the Pew report: "those students who were using the computer lab to do academic-related work made use of commercial search engines rather than university and library Web sites."[20] Data analyzed in Eric Novotny's 2004 study "I Don't Think I Click" further confirm this trend. For example, patrons frequently expect the library's catalog to function as an Internet search engine using natural language and keyword searching instead of some of the more advanced search techniques, such as using Boolean operators or synonyms.[21] Novotny also found that the novice users observed were quick to click and try many different links and searches, but their speed also extended to their willingness to quickly reject a particular strategy or resource that was not immediately satisfactory.[22] When these kinds of users, with a hair trigger for dumping a resource, come in contact with one that does not perform the way they expect, the result can be disastrous for libraries trying to win over new users.

However, it's not all bad news. If libraries had stuck to their paper-based resources, then there would be cause for serious alarm, but, as indicated from the previous discussion, the profession took cues from their patron communities long ago and began to create the infrastructure required in a high-level information environment. As the Pew study found, "University libraries have tried to adapt to the information resources that the Internet offers by wiring themselves for students' demands. For example, computers are scattered throughout libraries to allow students to search for resources easily."[23] Greater than recognizing the need to get wired has been the realization that students are coming to college with a different research skill set than before: "college students seem to rely on information seeking habits formed prior to arriving at college."[24] This finding is another example of how libraries have had to adapt to a generation of users with research habits that may not include any knowledge of or experience with library resources. With the knowledge that incoming classes will possess increased experience with online searching, it is incumbent upon librarians to recognize these skills as they select e-resources and design instruction.

In order to understand today's incoming freshman, libraries need to examine where these students feel comfortable obtaining research information. In Pew's 2001 "Teenage Life Online" report, the researchers discovered that there are still students whose primary access to Internet resources is through school (11 percent). However, both parents and teens agree that the Internet is a key resource for completing schoolwork, with 94 percent of teens surveyed using the Internet for school research. While the report does not indicate which resources these parents and teens are using (possibly they are local library resources available online), it is apparent that these early research behaviors are key factors for academic libraries to consider when choosing resources, designing interfaces, and instructing new users.[25]

The Internet sets up a brand new norm for users, one in which keyword searching and downloading are the main methods of accessing information. The challenge for libraries is taking advantage of the skill sets students are bringing to the table while at the same time educating them about the value of resource selection and evaluation.

The Online Computer Library Center (OCLC) report *Perceptions of Libraries and Information Resources* gives us some insight on how users perceive libraries as a whole—and e-journals specifically—in relation to the Internet at large. The bright side of the OCLC report for academic librarians is that college students consistently rate library usage higher than just about any other group in the survey. Indeed, the report states that "college students were more likely to indicate that their library usage has increased, at 44 percent."[26] In addition, even if it does not seem like it to the average reference or instruction librarian, college students appear to be aware of a library's e-resources. As the report indicates:

> College students are the most familiar with all the electronic resources and show a substantially higher use of electronic magazines/journals, online databases and electronic books. Library Web site usage is also highest among college students, at 61 percent. Both the US 14- to-17-year-old and 18- to 24-year-old segments indicate high use of the library Web site, at 44 percent each, but usage by other US age groups is low.[27]

While college students may be more aware of the online library resources available than other segments of the population, this does not mean that all of them are using these resources. For example, in this same study, students were asked to "please indicate if you have used the following electronic information sources, even if you have used them only once." Search engines came in with a whopping 82 percent, second only to e-mail. With Library Web sites at 61 percent, electronic magazines/journals at 58 percent, and online databases at 34 percent, these resources do not really seem to be endangering the supremacy of the search engine. The numbers become even more depressing when students explain where they typically begin their search for information. Eighty-nine percent started with a search engine, and library Web sites and online databases each pulled in a 2 percent share. Obviously, there is a lot that libraries can learn from search engines in terms of marketing and usability.[28]

So what's a library to do? Any businessperson would say "go with your strengths" and e-journals and databases happen to be one of those strengths. When asked to rate whether a particular resource "provides worthwhile information," 85 percent of college students either agreed or completely agreed that electronic magazines/journals provide worthwhile information, the highest score obtained by any library e-resource. Seventy-two percent of college students agreed or completely agreed that online databases provided worthwhile information as well.[29] These numbers show great awareness of library resources on the part of student users, but as the Pew study tells us, they still may not be users' first choice for information.

While the OCLC and Pew reports give us a bird's eye view on college student behavior, other segments of academic libraries' user communities need attention, namely faculty. While there may be less dependence on search engines among this crowd simply based on the many years that most faculty have survived without using them, the library literature suggests there is still strong evidence that the easier library tools become to use the more likely faculty will be to rely upon them. Early e-journal usage and acceptance research indicated that "the willingness of researchers to rely solely on e-journals is dependent, in part, on their successful use."[30] As computers have become more ubiquitous on campus and network speed has increased, one area that could still use improvement is interface design. Even though faculty may remember the older, slower methods of searching and locating articles, they may not be content with faster, yet oftentimes equally complex database interfaces.

Research also tells us that, for faculty, e-journal and database use is not always predicted by ease of use but more often by the type of research and discipline involved. Much of the e-journal usage research has focused on the sciences because of the strong response of STM publishers to move to the e-journal format. However, studies like Talja and Maula's 2003 article, which uses domain analysis, demonstrate that there are many factors that go into whether e-resources are widely used by a particular discipline. Talja and Maula assert that by breaking down areas further than "humanities" and "sciences," one can determine which forms of searching are most useful to a particular field and therefore be better able to interpret database and e-journal use data.[31] The authors conclude that, "although most fields we studied involve a mix of different search strategies: directed searching, browsing, chaining from seed documents, and sharing literature with colleagues, there were clear differences in the relative importance of these methods across fields."[32] Faculty often conduct research from a vastly different starting point than the average undergraduate. By studying both user groups, libraries and vendors can be more effective in designing appropriately multifaceted interfaces.

The unprecedented amount of information libraries have about their users has facilitated librarians' natural desire to improve the user experience and make library resources key to user success. What's more, it is the users themselves that can point us in the direction of future improvements. This is demonstrated by a twenty-one-year-old respondent's answer to OCLC's survey question, "if you could provide one piece of advice to your library, what would it be? 'Make a way to search through all of the databases with one search engine, instead of having to search each database individually.'"[33] It is easy to see that this student's description of federated searching technology (without even knowing that such a thing exists) is an indication that libraries are on the right path. There is no doubt that studies of user behavior have led us down this path of developing and implementing new technologies for information retrieval.

HOW DO LIBRARIANS AND VENDORS RESPOND?

One advantage that librarians have in facing the e-revolution is that they have no problem working on both large-scale international endeavors and small grassroots projects. In need of better technology to bring together the myriad e-resources available to users, libraries have turned to large-scale standards-based projects that allow for major vendor product development as well as homegrown solutions. Those same kinds of standards are also at work to make future use data collection more powerful than it has ever been. Meanwhile, libraries use these tools and this data to more carefully design their Web sites and resources while educating users beyond their humble keyword beginnings. In response to user data, libraries and vendors have much to say and offer in the form of linking, federated searching, user statistics, and user-centered design.

E-Helpers—Federated Searching and Linking Technologies

Users want simple, accurate, seamless electronic access to information: is that too much to ask? Increasingly, librarians are saying no, it is not. However, providing this access in the complex world of library resources is not so easy. Like the landscape of the United States Midwest, libraries are full of silos: discrete collections of information that do not play well with one another. Just as the consumers in a grocery store do not much care which silo their food comes from, most users are unconcerned by the database or archive of origin of the material they need. This is where federated searching and link resolvers enter the picture. They facilitate users' ability to move across silos easily and without any prior knowledge or understanding of the silos themselves. Indeed, with the introduction of standards like the OpenURL, libraries, publishers, and vendors are beginning to see the necessity of crossing the boundaries between vendors and publishers, even if they may not have seen the competitive advantage of interconnecting their resources initially. For more information on how these tools have taught the silos how to play ball, see Chapter 12, E-journal Management Tools.

User Statistics

Tools like federated search engines and link resolvers work because they are based on standards. Another area where the adoption of standards has worked to benefit the user experience is in the collection of user statistics. National Information Standards Organization's (NISO's) Standardized Usage Statistics Harvesting Initiative (SUSHI) and Counting Online Usage of Networked Electronic Resources (COUNTER) reports are a direct response to the concerns of early user studies involving usage statistics. The recognition of the power of usage statistics was immediate, but discord in comparing these statistics across vendor platforms left librarians at a loss for making definitive judgments. COUNTER offered a standardized framework for vendors to quantify the usage of online information resources and, as a consequence, provided libraries with access to the information needed to make cost-benefit analyses as well as outreach, collection development, and instruction decisions. COUNTER defines the types of statistics vendors will provide to libraries, an advantage that has unleashed the enormous potential of these statistics to benefit library users. COUNTER does not do it all alone, however; another standard, SUSHI, facilitates the delivery of COUNTER reports. Such partnerships have proven necessary to better synthesize the complexities of usage data. Carol Tenopir points out some of the problems with usage data in her September 1, 2005 *Library Journal* column stating "use data and cost per use can be misleading unless the size of the user population is considered. . . . Usage log amounts should be weighted and adjusted for the size of each subject population served."[34] Fortunately, companies like MPS Technologies have utilized these standards and introduced services like ScholarlyStats, which provides libraries with consolidated vendor usage statistics. Such initiatives demonstrate an even greater commitment to getting the most out of this collected data by decreasing the amount of administrative work needed for data collection and collation. This strong showing from the library and vendor community is further evidence of the importance of user feedback in libraries' efforts to enhance their e-resources.

Focus on Interface Design and User Education

Libraries are also partnering with patrons by studying user behavior to improve interface design. The challenge of good interface design for libraries is achieving a simple, easy-to-use interface without sacrificing the inherent power of the resource. Again, libraries benefit from the popularity of the Web, as there is no shortage of individuals studying how to reduce a Web site's learning curve to zero. Users want interfaces that are intuitive and that do not require learning new terminology or special techniques. The rise of giants like Yahoo!, Google, and Amazon demonstrate how users can embrace interfaces that appeal to their natural searching behaviors while at the same time providing them with a breadth of options for further navigation. One of the individuals working to teach designers how to build these kinds of intuitive interfaces is Jakob Nielsen. Nielsen has devised a variety of usability testing methods for fast and cheap feedback on interfaces. One such method is heuristic evaluation: "an informal inspection method in which evaluators assess whether an interface complies with recognized usability principles or heuristics."[35] Nielsen has popularized interface evaluation as a necessary part of development that does not require thousands of users to be valid. As he explained in a 2001 *EContent* article, "for judging the quality of a user experience, you absolutely have to do an observational study where you look at a small number of people in great detail and see how they use the products."[36] Taking a page from the world of anthropology and sociology (and even bringing experts in these fields on board), a variety of methods can be used to examine users in their natural habitat interacting with interfaces. Bryan Heidorn, Bharat Mehra, and Mary Lokhaiser used five different methods to collect data about the types of research that botanists conduct, how they conduct it and how they interact with various interfaces. The methods they used included "interviews, focus groups, immersions and field observations. The value of this multitiered approach to user research is clear: Each method can be used to compensate for the weaknesses of the other methods."[37] For an example of user-centered interface design research, see Chapter 8, "Enhancing E-resources by Studying Users."

Ideally, interfaces should be so well designed and intuitive that users do not need instruction on their use. Until then, issues related to user-interface interaction will have to remain on the list of topics addressed by library instruction programs. In addition, the reality for many libraries is that some segments of the user community, such as distance education students, will never enter the library building or even make it to the library Web site, much less an instruction session. The response to both scenarios has been the same: go to where the people are. Whether it is course management software or portals, RSS feeds or podcasting, chat, or good old e-mail, reference and instruction librarians have seen the future and its name is outreach. This concept of outreach extends far beyond university channels as well. With projects such as OpenWorldCat and Google Scholar's OpenURL linking, users are being actively drawn into library resources, sometimes without their knowledge. Once they are on the library's Web site, users are treated to customized links to their libraries' reference services and interlibrary loan. When full text is unavailable, librarians seize the opportunity to not only customize options for users to acquire the needed information but they can also provide further instruction on how to search more effectively and save users' time.

Barbara Macke exemplifies this new approach to user education that has been born out of the Google revolution when she comments that

> There is a difference between providing information and providing information in a palatable format, and this difference frequently involves the act of interpretation. In the undergraduate library, we may have the illusion that we are dealing with books, articles, and reference materials, all rich with information, but we are really dealing with the immediate and pressing needs of our students, and the accompanying need to tweak those information sources to make them understandable, more accessible, and in many cases, even appealing.[38]

Rather than seeing user education as an absolute necessity, librarians can approach it as a marketing challenge. They must demonstrate to users the value their products have over others. Online tutorials provide interactive environments for users where they can work at their own pace to acquire the skills they desire. Reference and instruction librarians are at a unique advantage of being able to educate patrons on the value and use of Internet resources they have grown accustomed to while also promoting the competitive advantages of library resources such as organization, authority control, and cost-effectiveness.

Librarians can also provide a much higher level of user assistance by understanding the ways users conceptualize knowledge and use information. Research focusing on these subjects is becoming more popular. For example, Clarence Maybee's study in the January 2006 *Journal of Academic Librarianship* examines a phenomenographical approach to looking at users' conception of information use. Maybee found that undergraduate students conceived of information use in three distinct ways: sources, processes, and knowledge. Maybee recognizes the value of understanding these concepts:

> A relational approach should be employed to embed information literacy values into course curriculum that focuses on students conceptualizing information use in increasingly complex ways. Knowing the three ways that undergraduates conceptualize information use will allow for the creation of user-centered information literacy pedagogy designed specifically to strengthen student learning.[39]

Thanks to such user-centered research studies, libraries are better prepared to design services and programming. In essence, the online environment not only offers the opportunity to provide the right resource at the right time, but also the right assistance at the right time. For example, tools such as link resolvers ensure that users efficiently move from databases to the library's licensed content. However, when users reach dead-ends or are unable to understand the options they are being given, librarians can offer customized instruction in the link resolver intermediary window that can offer options as well as connect users to tutorials or a real live person through chat reference. Today's library user is offered both unprecedented access to information resources as well as easy avenues to obtain assistance in all aspects of information retrieval.

CONCLUSION

Libraries are full of seams—just ask anyone who has had one or two additions put onto their library building. Meanwhile, librarians have perpetually strived to caulk and collocate their way to seamlessness for their users. E-journals and databases have introduced a new environment that users and libraries have embraced as convenient and time-saving. With this new environment come new challenges and opportunities that are still under construction. Linking and federated searching, user statistics and in-depth analysis of user behavior, outreach, and innovation that take into account user needs and preferences all demonstrate the commitment of libraries and the vendors who serve them to develop resources and services that respond to their needs. The study of user behavior is increasingly bolstered by the wide variety of ways that libraries can now obtain user feedback. The future may see increased use of Wiki-like feedback mechanisms that allow users to become even greater partners in the process of interface and resource design. Whatever future efforts occur to streamline and enrich the research process, it is certain that they are firmly rooted in the e-journal and e-index developments that have come before. The ever-growing backbone of electronic scholarly communication that exists today is a direct result of the efforts of librarians, publishers, and vendors that have worked over the last twenty years to envision new, efficient models for the discovery and delivery of information. Following their example, if library professionals want to know what the future of libraries will be, they need look no further than the patrons who use them and, maybe more importantly, those that do not.

NOTES

1. Brett Butler, "Scholarly Journals, Electronic Publishing, and Library Networks: From 1986 to 2000," *Serials Review* 12 (1986): 48.
2. Ibid.
3. Eric L. Morgan, "Description and Evaluation of the 'Mr. Serials' Process: Automatically Collecting, Organizing, Archiving, Indexing and Disseminating

Electronic Serials," *Serials Review* 21, no. 4 (1995), http://search.epnet.com .ezproxy.samford.edu/login.aspx?direct=true&db=buh&an=9604010689.

4. Ibid.

5. David J. Pullinger, "Learning From Putting Electronic Journals on Super JANET: The SuperJournal Project," *Interlending & Document Supply* 23, no. 1 (1995), http://proquest.umi.com.ezproxy.samford.edu/pqdweb?did=116350164& sid=1&Fmt=3&clientId=22446&RQT=309&VName=PQD.

6. John M. Budd and Lynn Silipigni Connaway, "University Faculty and Net-worked Information: Results of a Survey," *Journal of the American Society for Information Science* 48 (1997): 843-852.

7. Suely Gomes and Jack Meadows, "Perceptions of Electronic Journals in British Universities," *Journal of Scholarly Publishing* 29 (1998): 174-181.

8. Donnice Cochenour and Tom Moothart, "E-Journal Acceptance at Colorado State University: A Case Study," *Serials Review* 29 (2003): 17.

9. George S.Machovec, "Electronic Journal Market Overview—1997," *Serials Review* 23, no. 2 (1997): 31, http://search.epnet.com.ezproxy.samford.edu/login .aspx?direct=true&db=buh&an=9711262265.

10. Ibid., 2.

11. Carol Tenopir, Donald W. King, and Randy A. Hoffman, "Scientists' Use of Journals: Differences (and Similarities) Between Print and Electronic," in *National Online* 2001: Proceedings (Medford, NJ: Information Today, 2001), 469-481.

12. Carol Tenopir, Donald W. King, Carol Hansen Montgomery, and Sarah E. Aerni, "Patterns of Journal Use by Faculty at Three Diverse Universities," *D-Lib Magazine* 9, no. 10 (2003): http://www.dlib.org/dlib/october03/king/10king.html.

13. Thomas E. Nisonger, *Management of Serials in Libraries* (Englewood, CO: Libraries Unlimited, Inc., 1998), 160.

14. Butler, "Scholarly Journals," 50.

15. Carol Tenopir, Donald W. King, Peter Boyce, Matt Grayson, Yan Zhang, and Mercy Ebuen, "Patterns of Journal Use by Scientists Through Three Evolutionary Phases," *D-Lib Magazine* 9, no. 5 (2003): 12, http://www.dlib.org/dlib/may03/ king/05king.html.

16. Jie Tian and Sharon Wiles-Young, "The Convergence of User Needs, Collec-tion Building and the Electronic Publishing Market Place," *The Serials Librarian* 38 (2000): 333.

17. Judit Bar-Ilan, Blumma C. Peritz, and Yecheskel Wolman, "A Survey on the Use of Electronic Databases and Electronic Journals Accessed Through the Web by Academic Staff of Israeli Universities," *Journal of Academic Librarianship* 29 (2003): 353-354.

18. Steve Jones and Mary Madden, *The Internet Goes to College* (Washington, DC: Pew Internet & American Life Project, 2002), 2, http://www.pewinternet.org/ pdfs/PIP_College_Report.pdf.

19. Ibid., 12.

20. Ibid., 13.

21. Eric Novotny, "I don't think I click: A Protocol Analysis Study of Use of a Library Online Catalog in the Internet Age," *College and Research Libraries* 65,

no. (2004): 525-537, http://www.ala.org/ala/acrl/acrlpubs/crljournal/crl2004/november/Novotny.pdf.

22. Ibid., 530.

23. Jones and Madden, *The Internet Goes to College,* 13.

24. Ibid., 13.

25. Amanda Lenhart, Lee Rainie, and Oliver Lewis, *Teenage Life Online* (Washington, DC: Pew Internet & American Life Project, 2001), 5, http://www.pewinternet.org/pdfs/PIP_Teens_Report.pdf.

26. DeRosa et. al., *Perceptions of Libraries and Information Resources* (Dublin, OH: OCLC Online Computer Library Center, Inc, 2005), 1-4, http://www.oclc.org/reports/pdfs/Percept_all.pdf.

27. Ibid., 1-11.

28. Ibid., 1-17.

29. Ibid., 1-33.

30. Donnice Cochenour and Tom Moothart, "E-Journal Acceptance at Colorado State University: A Case Study," *Serials Review* 29 (2003): 17.

31. Sanna Talja and Hanni Maula, "Reasons for the Use and Non-Use of Electronic Journals and Databases: A Domain Analytic Study in Four Scholarly Disciplines," *Journal of Documentation* 59 (2003): 686.

32. Ibid., 685.

33. DeRosa et al., *Perceptions of Libraries,* 1-19.

34. Carol Tenopir, "Inundated With Data," *Library Journal,* Sept. 1 (2001): 31.

35. Darlene Fichter, "Heuristic and Cognitive Walk-Through Evaluations," *ONLINE* 28 (2004): 53.

36. Thomas Pack, "Use It or Lose it: Jakob Nielsen Champions Content Usability," *EContent* 24 (2001): 46.

37. P. Bryan Heidorn, Bharat Mehra, and Mary F. Lokhaiser, "Complementary User-center Methodologies for Information Seeking and Use: System's Design in the Biological Information Browsing Environment (BIBE)," *Journal of the American Society for Information Science and Technology* 53 (2002): 1257.

38. Barbara Macke, "Roaches, Guerillas and Librarians on the Loose," *Journal of Academic Librarianship* 31 (2005): 586-587.

39. Clarence Maybee, "Undergraduate Perceptions of Information Use: The Basis for Creating User-centered Student Information Literacy Instruction," *Journal of Academic Librarianship* 32 (2006): 84.

Chapter 8

Enhancing E-Resources by Studying Users: The University of Rochester's Analysis of Faculty Perspectives on an Institutional Repository

Nancy Fried Foster
David Lindahl

INTRODUCTION

As libraries have increased their use of such digital materials as e-journals, electronic theses and dissertations, and gray literature stored in digital repositories, they have sought new and better ways to make these materials findable and usable by students, faculty members, and other patrons. This has meant developing online tools, in addition to the library catalog, that make it possible to search large collections of e-resources, sort through and pare down long lists of "hits" to find the most appropriate ones, and then gain access to these materials in only a few "clicks." It might seem that the requirements for these online tools would be so obvious that developers would never have to consult with librarians, let alone users, to build them successfully. However, experience shows that the best online tools depend on participation throughout the design and development process by a variety of people, including students and faculty members, librarians and library staff, interface designers, and software engineers.

In this chapter, we describe a design project that demonstrates how user input can lead to improved design and more effective use of e-resources. In this project, the University of Rochester's River Campus Libraries collected information about faculty work practices in order to document obvious needs and reveal less apparent ones related to our new institutional repository. The mapping process we used to link raw data to requirements, and from there to specifications and solutions, provided a heuristic for problem solving and a structure for ensuring that our repository really did meet the needs of our faculty.

MAKING OUR INSTITUTIONAL REPOSITORY WORK

An institutional repository (IR) is an electronic system that captures, stores, and provides access to the digital work products of a community. Its main components are a repository of content, associated metadata, and a user interface. In the case of a university IR, success means meeting both institutional and individual needs, filling the repository with searched and cited scholarly work of enduring value, serving as a showcase of the intellectual output of the university, and enhancing scholarly communication.[1]

In some ways, an IR is a simple extension of a university's library. Both, collect and preserve scholarship (the content), organize it (using metadata), and make it available to members of the university (through various interfaces).

However, adding an IR to a library's suite of collections and services is not so simple. This has partly to do with the fact that libraries are currently in transition, having changed from the days of card catalogs, paper indexes, print journals, and primarily paper collections. Some digital tools that started as time-saving options are now required components of library services. The replacement of card catalogs with online catalogs took years but is now complete in many libraries. However, the transition from legacy to digital systems is far from complete, and the library still houses vast amounts of physical material. The library model offers metaphors for IR design but does not serve as the IR's model. In addition, connecting the IR to the library is not straightforward, due to the differences between them.

Libraries traditionally brought in works primarily from outside the university, mainly on paper, and made them available to faculty members. The academic library's role was to buy the end products of scholarly conversation, that is, books and journals. Now, libraries are also taking responsibility for making the work of the university's own faculty available to an audience outside the university, usually in digital format. As they do this, libraries enter into the live, prepublication conversation as never before. IRs support this by providing a way to store, organize, find, and deliver the university's scholarly output, not just to members of the university but to anyone in the world with an Internet connection.

While libraries build their collections mainly through purchases of commercial publications, IRs are built by faculty members submitting their own work. In other words, librarians are dependent on the people who generate the content to make the deposits into the IRs, since they cannot just purchase the content as they would a published book. Indeed, the library could build an IR and then find that none of the faculty members deposit their content into it. In fact, this has been happening: faculty members have not rushed to take advantage of the new IRs that have been set up to preserve and disseminate their work.[2]

In the process of conducting research on prospective faculty users of our university's IR, our mapping technique has helped us understand some of the complex reasons why faculty members have been so slow to put their work into established university IRs. These reasons include the mismatch between the language of librarians and the language of faculty members and the complex relationship among the needs of faculty members, librarians, university administrators, and others. Our way forward, as IR implementers, is to create an enhanced user experience on top of our existing IR system that is a better match with the expressed needs of all users. That is, our path is to understand what users really need to do and then to design e-resources to meet those needs.

WHAT USERS REALLY WANT

We began implementing an IR at the University of Rochester by selecting DSpace as our platform.[3] Initiated through a partnership between the Massachusetts Institute of Technology Libraries and Hewlett-Packard, DSpace is an IR application that is being developed collaboratively by a large group of committed users. DSpace is just one open-source collaboration among big universities; another well-known one is Fedora.[4] We chose DSpace because it offered us an opportunity to be part of the development process and bring our user-centered design methodology to bear on the emerging product.

Once we had completed initial installation, one of our librarians began to pitch the IR to faculty members using typical DSpace promotional language, shown in Figure 8.1.

Meanwhile, we began a user study to collect the information we needed to customize the interface and enhance the IR so that it would be more usable and interesting to our faculty.[5] We thought the best way to collect this information would be by means of a work-practice study that is a fine-grained, on-site study of individual researchers engaged in typical research practices.[6] We took a team approach, involving a computer scientist, anthropologist, software engineer, graphic designer, and public service and catalog librarians. This team videotaped over two dozen faculty members in their work settings performing research tasks and showing us how they organize their physical and cyber workspaces.

We did not simply ask faculty what they wanted in terms of an institutional repository. Instead, we tried to ask questions about their research and about scholarly communication in general. We also probed their use of digital tools in their work and their use of the library.

We transcribed all observational sessions and analyzed the transcripts and video to yield a range of findings, from lists of faculty needs

---

DSpace Features and Benefits

- Large-scale, stable, managed long-term storage
- Support for a range of digital formats
- Visibility for research results
- Persistent network identifiers
- Flexible and simple submission process
- Search and delivery interface
- Digital preservation services

FIGURE 8.1. Typical Promotional Language

and storyboards of the faculty research process to models of faculty preferences and perspectives. For example, we learned that

- Faculty members are passionate about their research;
- They want digital tools that work but they do not care how they work;
- Placing their work in an IR is only of value to faculty members if other scholars find it there, use it, and cite it.

As Table 8.1 shows, faculty members have a very different perspective on DSpace than other users and stakeholders. Indeed, not just faculty members, but all types of users and stakeholders have their own distinctive and different needs and perspectives.

Sometimes, these different interests have led to disabling mismatches between IRs and users. The more we learned about our IR, the more we realized that it was organized to suit the purposes of nonfaculty users and was described in nonfaculty language. Owing to this, most faculty members looking at the IR failed to see how it could benefit them. Here are a few specifics,

- The language of the IR, shown in Figure 8.1, did not correspond to the language that faculty members used in our interviews with them. For example, while we saw that almost all faculty members had problems with broken links as they searched for resources on the Web, only a few mentioned this as a problem, and only one used a phrase that even approached "persistent network identifiers." This was a relatively easy problem to correct, but until we learned how to talk to faculty in their own terms, we missed many opportunities to recruit them to IR use.
- The organization of DSpace into departmental communities and subject area collections does not correspond to the spontaneous, shifting networks of scholars engaged in related projects. DSpace is organized to present the content of the institution as a whole and thus to meet institutional needs. We think we have found a partial solution. In addition to the community/collection hierarchy of DSpace, we are building "Researcher Pages" (Figure 8.2) so that faculty can organize their own work on searchable pages and include links to the sites of other people in their networks.

TABLE 8.1. Examples of IR Users and Their Different Interests

Type of User or Stakeholder	Main Interests in Institutional Repository
University administrators	Showcase university's scholarly output Contain the expense of scholarly journals
Individual faculty members	Find resources to use in one's work Keep one's work secure and accessible Disseminate one's work
Software developers	Make IR work Make good on all claims of IR functionality
Librarians	Develop, catalog, and preserve collections Help faculty and students find resources

FIGURE 8.2. Researcher Page Enhancement. *Source:* This sample Researcher Page includes a photo and an area for contact information and professional interests. On the left side, faculty can present their work using a hierarchy of folders and include links to other sites and related content. Reprinted with permission.

In order to ensure that faculty members invest their time and energy in using an IR, we must both speak their language and directly address their needs with the technology. We are using our mapping technique to translate their needs into technical specifications and ensure that we fully support those solutions.

COMPLEXITIES OF IMPLEMENTATION

There is a dilemma with DSpace. Since it addresses institutional requirements so well, institutions across the country have deployed this IR. However, because of the mismatches described in the previous section, faculty members are not putting as much content into DSpace as the institutions might have expected.

Further complicating the picture, our research on faculty members has revealed a range of needs for Web-based services to support research activities, some of which are supported by DSpace and others are not, as shown in Table 8.2.

The simple listing of needs revealed that, while the IR was meeting some faculty needs, it was not necessarily supporting their greatest needs or those that came earlier in the research process. That is, faculty members need to do their writing, alone or with coauthors, often going through many revisions, before they are ready to archive or publish it. We realized that the IR would be a greater success if we offered faculty members a system for managing the authoring process, especially if it easily led into the self-archiving and self-publishing features of the IR.

We also started to understand why even faculty members whose completed work seemed perfect for the IR were not depositing it. One of the reasons was that they did not have reason to believe that anyone else would find, use, or cite it. The solution to this problem, we believed, would be to tightly couple the act of depositing content into the repository with the rewards of presenting that work to colleagues. We designed our Researcher Page to do this by providing faculty members with a single, user-friendly interface for depositing and showcasing their work. We believe that this enhancement, which attends to the faculty need to present their own work to their colleagues, will motivate faculty to use DSpace and deposit their work.

As we identified and examined more and more of these needs, we discovered that we could trace out the way they did or did not connect to specific bullets of the DSpace promotional language (Figure 8.1). In many cases, faculty needs did map to DSpace features, but DSpace features were only a partial solution to the faculty need. In other cases, the DSpace promotional bullets referred to DSpace features that the product alone could not deliver.

TABLE 8.2. Faculty Needs and Technology Solutions

Faculty Need	Technology Enabler
Make their own work available to others	DSpace
Preserve digital items	DSpace
Ensure that documents are persistently viewable or usable	DSpace
Have someone else take responsibility for the server	DSpace
Make digital items permanently accessible	DSpace
Work with coauthors	Document authoring and versioning system
Work from different computers and locations, both Mac and PC	Web-based document authoring and versioning system
Have easy access to other people's work	Web-based repository organized by the researcher with search tools and permission control
Keep up in their fields	Improved search tools and general adoption of DSpace to improve the accessibility of all scholarship
Organize their materials according to their own scheme	Individual control of navigation structure and permissions for Web access of their work
Control ownership, security, and access	DSpace and document management permissions with adequate granularity and enhanced user interface
Be sure not to violate copyright issues	DSpace enhancements and librarian support
Keep everything related to computers easy and flawless	User-centered design
Reduce chaos or at least not add to it	User-centered design
Not be any busier	User-centered design

At issue is the importance of collecting and then rigorously mining the data in order to bring all needs—not just obvious needs—to the surface. Once we understand the full range of needs, we can design or retrofit e-resources to meet those needs. Furthermore, we want to be sure both that our solutions meet real needs and that our solutions are combined with necessary services and supports and are configured to be efficient and effective. Thus, we need a tool to map from user needs to solutions and back again.

ADDING MAPPING TO OUR TOOLKIT

We have worked with so many faculty members and collected so much data that we have developed a need for a tool to manage, represent, and work with the data. The tool that we have used most successfully is a simple set of maps that connect all the data and ideas around each category of user needs that we find.

Each map traces a logical path from user-stated needs to fully supported technology solutions. We make one map for each general kind of user need, and we include a set of user quotes in the map to maintain access to the users' own language. We analyze these quotes and distill a short list of needs that are now stated in the language we use in the project. After that, we brainstorm and define the system specifications that we think will address those needs and the technology solutions that could bring those enablers to our users. Finally, we map out all the forms of support that our technology solutions would require in order to work properly.

In order to understand how our mapping technique works, it helps to consider the relationships that exist among people and the connections that people have to their tools and objects. In our current example, we are most interested in faculty relationships and connections, as shown in Tables 8.3 and 8.4.

University faculty work within an ongoing scholarly conversation, engaging with colleagues, students, librarians, and others in order to learn more and say more about the topics and issues that matter to them. If our IR can provide digital tools in support of faculty relationships and work connections, then our IR will succeed. At the same time, the goals of librarianship—collection, preservation, organization, and access—will be achieved. Our mapping exercises are guiding us

TABLE 8.3. Some Faculty Relationships

Relationship	Examples
Faculty ←→ colleagues	Collaborate, share data
Faculty ←→ librarians	Get research and teaching support
Faculty ←→ departmental and institutional administrators	Develop curriculum, provide service to the institution, advance
Faculty ←→ students	Teach, advise

TABLE 8.4. Some Faculty Connections

Connection	Examples
Faculty ⟷ artifacts of topic	Collect, create, and study quilts, audio tapes, images, code, fieldnotes, and so on
Faculty ⟷ data	Collect, share, generate, and manipulate data
Faculty ⟷ repositories	Disseminate own work quickly, for example, through arXiv.org; get feedback and credit for original work
Faculty ⟷ journals, books	Share own work; learn about the work of others

to technical and process solutions that will support our faculty in just this way. In the following section, we give an example of one of our maps to demonstrate the power of this simple tool.

AN EXAMPLE OF MAPPING

In this section, we follow a thin path from raw data (a few actual user statements), through analysis, to solutions. We provide this example in order to illustrate our mapping process in use. We hope to show how a particular e-resource, our repository, benefits from collection, analysis, and use of information about the people who will actually use it.

We were initially motivated to map user needs to DSpace features to see if the exercise would reveal the reasons why faculty were not interested in the IR. In this example, we will show part of that mapping by focusing on a portion of the user needs that are related to finding archived material. As we work through this example, we will be looking at one particular DSpace feature—persistent network identifiers—to see where it emerges in the mapping process, that is, to see how closely this feature maps to articulated user needs.

We were also concerned about backing up the claims we made for DSpace. That is, when we say that DSpace "offers persistent network identifiers," what really makes that happen? Users might assume that persistent network identifiers are somehow part of DSpace, that they are just coded into the IR in some way. The systems analyst and software engineer know that this is not true, of course. However, even our technical people find it essential to map out all the supports that are necessary for a system to work or for a feature to be offered in good faith. In our example, therefore, and by way of illustration, we trace out the necessary supports related to persistent network identifiers.

In addition, the more we learned about faculty needs, the more we wanted to create the big system picture of how we could meet those needs. Indeed, we began to see that meeting even a few of the most important faculty needs would likely increase faculty use of the IR. Mapping was thus our tool for identifying the specifications of the whole system. Here, we include the specifications related only to our small, focused example.

When we use our mapping tool, we start with the transcripts of sessions in which users have talked about what they are doing as they use their digital tools or conduct research. We extract user statements that relate to a particular category of need, usually having several pages of excerpts for each theme. In our example, we start with four brief excerpts from four separate interviews for illustrative purposes (Figure 8.3).

Once we have compiled user statements, our anthropologist analyzes them and works with an interdisciplinary team (computer scientist, software engineer, graphic designer, librarian) to identify the

---

User Statements on Finding Archived Material

User #1: My first class this semester. [Clicking, typing:] I need to open that. Um, my first class this semester I did a PowerPoint presentation of pieces [clicking mouse] that I had not done that before, so I've taken [searching for document]

User #2: Yeah, but it's not going to work. [Looking for document, extensive mumbling, typing] Let me think for a second. [Mumbling, typing] I think I had it, I ... [Trails off]

User #3: If DSpace were really there, really preserving the URLs, then we might choose to set these links to points in DSpace, and that way we could actually have our server be mobile and changeable and yet have the URLs in DSpace be fixed.

User #4: Addresses would be fixed, right, so then I could say that I put a data file in there, so then people could go to DSpace right from my webpage if I had that URL in there.

FIGURE 8.3. Mapping Example Step 1

underlying needs. Figure 8.4 lists the analyzed needs related to finding archived material. We use IT language when we list the analyzed needs that underlie the users' own statements.

We can see that the needs that underlie faculty statements extend far into faculty webs of relationships and connections. Meeting these needs will enable professional relationships; that is, they will support the context of scholarly work. By attending to these relationships, we build not just a usable interface but also a social interface that supports cooperative work. Figure 8.5 shows a few of the relationships and connections that are relevant in our example.

Next, the interdisciplinary team brainstorms a list of everything it would take to meet the analyzed needs. Normally, a map would include everything that relates to all analyzed needs. However, in this example, we will treat only one need: "make digital items permanently accessible." Figure 8.6 lists all the specifications we have identified in relation to this need.

In the next step, the system designer leads the team through a rigorous comparison of products and selects those that meet system specifications. Again, in our example, we are only following the threads related to "making digital items permanently accessible" (see Figure 8.7).

Analyzed User Needs for Finding Archived Material

- Make own work available to others
- Have someone else take responsibility for the server
- Make digital items permanently accessible
- Have easy access to other people's work
- Reduce chaos or at least not add to it

FIGURE 8.4. Mapping Example Step 2a

Affected User Relationships and
Connections for Finding Archived Material

Faculty ⟷ colleagues
Faculty ⟷ librarians
Faculty ⟷ artifacts of topic
Faculty ⟷ data

FIGURE 8.5. Mapping Example Step 2b

System Specifications for Making Digital Items Permanently Accessible

- Provide a physical storage system that is managed to include standard operating procedures such as backup, mirroring, refreshing media, and disaster recovery
- Ensure that a storage area will remain available and stable forever
- Assign a globally unique identifier to each item that will never change
- Provide access to items via the Web
- Provide a mechanism to indicate if files have been modified
- Provide a mechanism to ensure that specific file formats can be rendered the same way in the future as they can be rendered today

FIGURE 8.6. Mapping Example Step 3

Technology Solution(s) for Making Digital Items Permanently Accessible

- DSpace
- Persistent Network Identifier Service (CNRI Handle System)
- Internet

FIGURE 8.7. Mapping Example Step 4

Finally, our computer scientist, software engineer, and systems people drill down to identify everything it would take to make our technology solution(s) work. At this stage, we have left the realm of pure technology. Figure 8.8 includes hardware and software, of course, as well as the full range of resources required to make even the best hardware and software work. In other words, this is the place to include the dollars, person-hours, expertise, and procedural care that support system success.

In this mapping example, we wanted to see how "persistent network identifiers," one of the bulleted DSpace features, related to user needs. To do this, we explored part of a category of user needs we called "making digital items permanently accessible." In doing so, we gained greater understanding of our faculty's lack of interest in the IR. Conversely, we discovered something about what it would take to make this e-resource workable and valuable to faculty members.

Necessary Supports for Making Digital Items Permanently Accessible

DSpace

- Software platform for DSpace (UNIX OS, Java 1.4, Apache Ant 1.5, Jakarta Tomcat 4.x/5.x)
- Database software (PostgreSQL 7.3)
- Server hardware
- IT support organization (backup, mirroring, refreshing media, and disaster recovery)
- Institutional commitment to long-term support of DSpace
- Investment in support for conversion or rendering of obsolete file formats

Persistent Network Identifier Service (CNRI Handle System)

- Depends on external entities

Internet
- Depends on external entities

FIGURE 8.8. Mapping Example Step 5

Our map shows that persistent network identifiers do not emerge until step 4, deep into the map and nowhere near user needs, whether stated in user or technical terms. In other words, we are touting a DSpace "feature" that does not capture the attention of users.

This particular example points to the more general problem that arises in implementing archives and other e-resources when we fail to differentiate among user needs, system features, system specifications, and necessary supports. Persistent network identifiers are a necessary support, dependent on other providers to make them work. When we differentiate between a user need and a necessary support, we can speak to real user needs in user language (see Figure 8.9). Differentiating among needs, features, specifications, and supports also allows us to develop a list of system specifications that covers the full range of user needs, or that portion of user needs that we can realistically address at one time. It not only helps us see the difference between IR features that are represented as part of the software design but also those that are really dependent on careful maintenance, ongoing enhancement and innovation, and permanent institutional resource commitments. Finally, mapping provides us with a fuller view of everything it will really take to make the system work, including services that are provided from outside the system itself.

Top Revised DSpace Features and Benefits

- This is all about your research: storing it safely and sharing it if you want to
- You can store items in DSpace permanently
- Your archived stuff is searchable through Google
- The DSpace submission process is far easier than posting documents to your personal or departmental website and you don't have to worry about backups
- You can give colleagues a URL to your item that will always work

---

FIGURE 8.9. Revised List of DSpace Features and Benefits

For our own purposes, mapping has helped us see how to move beyond our Researcher Page to the design and development of an authoring environment for IR users. We conceive a large system, of which the IR is one part, to support the full life cycle of the faculty writing and publishing process. This system will support the professional research context and the ongoing scholarly conversation, making it easier for our faculty to share their work with coauthors at the writing stage and with all others upon completion.

CONCLUSION

In our work, we have discovered how important it is to observe actual user behavior and align our technology to genuine user needs. We have developed a tool to help us do this by mapping user needs to specifications, technology solutions, and necessary supports.

While this chapter revolves around a particular application of work-practice study, participatory design, and mapping, we see the value of user research and mapping beyond this one case. When improving metasearch or other e-resource-related tools, research on faculty work practices would be invaluable. For example, fine-grained observation and interviewing of faculty members as they conduct research may reveal such practices as tasking students or assistants with this work, searching personal Web pages for citations, or a preference for certain types of databases. Such studies will certainly reveal the "workarounds" that faculty members use and that reveal both the limitations of their current tools and the ways that they would prefer to work. Actual research findings will point to particular solutions that might not have been anticipated and that will certainly work better than tools based on unquestioned assumptions.

Mapping has also shown us a pathway forward for the library as its use of e-resources continues to increase. Digital technology is now necessary for cataloging the library's resources and then finding them again, and it is often the preferred way to use those resources. Furthermore, digital systems are now preferred for archiving and publishing scholarly work. As digital technologies become even more important in scholarly work and communication, libraries will play a greater role in providing for a fuller range of faculty technology needs. We believe that identifying the needs through work-practice studies and then mapping needs to a range of potential solutions will increase our success in meeting them.

NOTES

1. Nancy Fried Foster and Susan Gibbons, "Understanding faculty to improve content recruitment for institutional repositories," *D-Lib Magazine* 11, no. 1 (2005), http://www.dib.org/dlib/january05/foster/01foster.html (accessed May 17, 2006).

2. Andrea Foster, "Papers Wanted: Online Archives Run by Universities Struggle to Attract Material," *Chronicle of Higher Education* 50, no. 42 (2004): A37; Morag Mackie, "Filling Institutional Repositories: Practical Strategies from the DAEDALUS Project," *Ariadne* 39 (2004), http://www.ariadne.ac.uk/issue39/mackie/intro.html (accessed May 17, 2006).

3. Massachusetts Institute of Technology, "Introducing DSpace," About DSpace, http://dspace.org/introduction/index.html (accessed June 11, 2004).

4. University of Virginia Library and Cornell University, The Fedora Project, http://www.fedora.info/ (accessed September 12, 2004).

5. David Lindahl and Nancy Fried Foster, "Use a Shoehorn or Design a Better Shoe: Co-Design of a University Repository," paper presented at the Participatory Design Conference, Toronto, ON, Canada (July 29-31, 2004), http://hdl.handle.net/1802/1384 (accessed May 17, 2006).

6. J.Blomberg, J. Giacomi, A. Mosher, and P. Swenton-Wall, "Ethnographic Field Methods and Their Relation to Design," in *Participatory Design: Principles and Practices,* eds. D. Schuler and A. Namioka (Mahwah, NJ: Lawrence Erlbaum Associates, 1993); F. Brun-Cottan and P. Wall, "Using Video to Re-Present the User," *Communications of the ACM* 385, no. 5 (1995), 61-71; M. Dawson, "Anthropology and Industrial Design: A Voice from the Front Lines," in *Creating Breakthrough Ideas: The Collaboration of Anthropologists and Designers in the Product Development Industry,* eds. S. Squires and Bryan Byrne (Westport, Conn.: Bergin & Garvey, 2002).

PART III:
EVOLVING TOOLS

Chapter 9

The Role of the Online Catalog As an E-Resource Access and Management Tool

Charley Pennell

INTRODUCTION

Reports of the death of the online catalog, while occurring with ever-increasing frequency and urgency, have been greatly exaggerated. Although there have always been problems in finding and interpreting serial records in the catalog, these problems have undoubtedly gotten worse in the past ten years as we have experimented with various ways of promoting remote electronic publications through our Online Public Access Catalogs (OPACs). This chapter will explore the present nature of the catalog, criticisms of its underlying structure and contents, add-ons and spin-offs that seek to redress some of its shortcomings, and specific problems for serial retrieval in a tool dominated by monographs. A review of the catalogs of thirty top-ranked members of the Association of Research Libraries (ARL) reveals the diversity of approaches that exist in the marketplace. It also shows that, while the majority of libraries now provide finding aids outside of the catalog—principally A-Z lists of e-journals and databases—the catalog continues to play a key role in the identification of a library's serial resources, both print and electronic.

IS THE CATALOG DEAD?

Over the past ten years, there has been no dearth of criticism of the online catalog, of the tools of cataloging ("MARC is dead"), and indeed of the art and science of cataloging itself. Since the rise of the Web in the mid-1990s, criticism of the catalog has intensified in almost direct proportion to the growth in and sophistication of the large Web search engines, most notably Yahoo, Alta Vista, and Google. Google, in particular, has raised users' expectations for simplicity of the user interface, speed of retrieval, and, most importantly, direct access to the actual text, image, or map described in retrieved citations. Recent tools like Online Computer Library Center's (OCLC's) Open WorldCat (http://worldcat.org), which exposes the world's largest bibliographic database to Web search engines, and Google Books (http://books.google.com), the massive research collection monograph digitization project, have caused many to question the relevancy of the local catalog, and by association, the local library facility itself.

Those who have publicly lambasted the OPAC note problems both in the MARC data feeding the catalog and in the underlying applications that Integrated Library System (ILS) vendors have written to serve up that data. Perhaps the best known characterization regarding the current state of the catalog and its underlying ILS is attributable to Andrew Pace, who has referred to the OPAC in recent presentations as "lipstick on a pig."[1] This sentiment has been echoed by Roy Tennant, Marshall Breeding, and Karen Calhoun, among others and has led to the creation of several quite active online forums for discussing the future of the catalog, most notably Eric Lease Morgan's e-mail list, *Next Generation Catalogs for Libraries* (NGC4LIB@listserv.nd.edu), the University of Rochester's *eXtensible Catalog (XC)* blog (http://extensiblecatalog.info/), and Karen Coyle's *futurelib* wiki (http://futurelib.pbwiki.com/).

The MARC communications format, now nearing its fortieth year in existence, is a frequent target for criticism from some of the same names noted in previous text.[2] Designed to minimize storage requirements and maximize data accessibility in an earlier era of relatively low processing power and expensive, inefficient storage, MARC has survived largely because it serves a niche marketplace (libraries) well and is supported by widely accepted and rigorously maintained standards. Central to both the promise and the problem of MARC is the heavy investment made to date by libraries, professional associations, ILS vendors, and service suppliers (bibliographic utilities, material vendors, authority record processors, etc.) in this standard. The promise is one of smooth data migration when the moment for change arrives, for MARC is still a highly granular description standard. The problem lies in our failure, due to a lack of capital, technical expertise, and perhaps will, to take the next step in abandoning a tool we have come to know so intimately.

Suggestions to open up the MARC standard have been largely based on a desire to encourage broader participation in the generation of bibliographic metadata and to then expose this data for harvesting outside of the immediate library community. To this end, there have been numerous attempts to simplify or even replace MARC and to create a more user-friendly data entry and editing interface. This would allow for the swift creation of bibliographic metadata with less investment in training and supervision and using lower-level staff, possibly even the content providers themselves. Abandoning the proprietary MARC data structure would allow libraries to break free of the constraints of the ILS marketplace, allowing them to use the presumably cheaper, more flexible, and more widely available tools and services already in place for the semantic Web. To meet this goal, a number of alternative SGML and XML DTDs and schemas have been proposed since the mid-1990s, including the minimalist Dublin Core element set and the more comprehensive (but MARC-based) MARC-XML and MODS schemas. Perhaps the most critical argument against MARC and other library bibliographic standards (RDA, AACR2, ISBD, etc.) is the need for greater extensibility in descriptive practice without the cumbersome and time-consuming draft/commentary process in use now through a voluntary committee structure.

Ten years after the introduction of Dublin Core, MARC-XML, and MODS, there is relatively little penetration of the traditional library catalog marketplace by XML schemas, though some vendors claim their ILS systems are able to handle one or more of them. Instead, a new generation of applications, the meta- or federated search engines like MuseGlobal and Ex Libris' MetaLib, have been added on top of the library automation mix, searching across both XML and MARC data, along with reference databases and other information stores. Much of the XML data feeding these metasearch tools is coming from areas not previously served by the catalog, and whose storage and retrieval needs are not a good fit for MARC, namely archives/special collections, Geographic Information Systems (GIS), and digital text and image repositories. This relegates the legacy ILS, upon which our catalogs rest, to the domain of MARC.

It is important to remember, however, that our systems only communicate (display to the cataloger, export and import) in the MARC format. Internally, the ILS stores data in a structure more amenable to manipulation by the system and to linking with nonbibliographic data, such as authority files, holdings, acquisition, and circulation system data. Obviously, security concerns would mean that further exposing bibliographic data could only be done if sensitive personal and financial data were partitioned behind firewalls or were to be held on a different server altogether. To this end, libraries are reexploring an earlier system architecture in which the inventory functions of the ILS are isolated from public searching. This model originated with computer-output microform (COM) catalogs and serial union lists in the 1970s and was transferred to CD-ROMs in the mid-1980s with products like The Library Corporation's *Intelligent Catalog* and GRC's *LaserQuest.* Data from the library's catalog was extracted quarterly, monthly, or weekly and sent to a vendor for mastering onto the presentation media. The end user of the resulting catalog was often forced to consult a cumulation (compiled annually or semiannually) plus supplements to find reasonably current information, and of course, the availability or circulation status of the title was impossible to ascertain. A return to this model, albeit on a vastly improved time cycle, is evident in products like North Carolina State University (NCSU) Libraries' implementation of Endeca, which attempts to satisfy user expectations conditioned by the speed and simplicity of Google.[3] Endeca, which was developed for Web-based commercial sales databases, relies on periodic snapshots of the library's bibliographic file, which is indexed and served up from the host server's memory to enable lightning fast retrieval, even from catalogs holding several million titles. Endeca exists parallel to the ILS with its sole justification the quick satisfaction of user search requests.

THE ROLE OF THE CATALOG
IN IDENTIFYING SERIALS

Complaints about the OPAC as a resource discovery tool, particularly for identifying materials on subjects of interest, are well summarized in Antelman, Lynema, and Pace's article cited in previous text,[4] but true resource discovery for periodicals does not happen in the OPAC. Instead, it occurs in periodical indexes and full-text databases where the more relevant author and subject terminology can be searched, namely within the articles themselves. Classic user behavior is to approach the catalog with a citation in hand and a desire to know if the library owns, or has electronic access to, the needed issue of a particular title. The rise of full-text databases and OpenURL linking software like SFX has obviously reduced the need for the intermediary services of the catalog, at least for titles available electronically. Nevertheless, there are many patrons who still need to determine whether a library has a given volume or issue of a periodical, where it is located (including those on the Web), and if it is available at that precise moment. The catalog often fails to deliver even this basic information.

Periodical searching in the OPAC is often for a known, or at least partially known, title. One might expect this to be a simple task, but often it is not. As a percentage of the total number of titles in a typical library catalog, serials are vastly outnumbered by monographs. For example, in the NCSU Libraries' catalog serials represent only 2.66 percent of all titles, compared with 72.83 percent for books and another 9.15 percent for e-books; statistics on the contents of the NCSU catalog from October 1, 2006 are posted at: http://www.lib.ncsu.edu/cataloging/stats/Snapshot1Oct06.html. Many periodicals have generic or nondistinctive titles that become lost in a sea of monographic titles, especially when a patron approaches the catalog through the common "quick search" interfaces that default to a general keyword or keyword-in-title search.

Even with title browse, it is sometimes a tedious effort to get to titles such as *New York Times, Scientific American,* the *Economist,* or *Time,* all of which frequently appear as part of monograph titles or in uniform title added entries; a fuller discussion of this problem can be found in the *Serials Librarian* article "Magnifying the ILS with Endeca."[5] OPAC designers have tried to solve this problem by enabling search limits by format, either in the form of a "periodical/journal title" search, which applies limits behind the scene, or a drop-down list of format or item types from which the user can select terminology like "Serial" or "Periodical/Newspaper." This helps for most periodical titles, though not for some in which the desired title phrase occurs frequently. The addition of separate records describing microform and electronic versions of periodicals, and even multiple manifestations of electronic versions, plus the addition of duplicate records describing consortium members' holdings, also adds to the clutter in the catalog index display.

Once a user finally identifies the desired title in the catalog, this user must attempt to interpret the data found in the display. The screen layout of many OPACs can be extremely confusing, with description, holdings, and links to electronic versions as well as to earlier/later titles, seemingly dropped at random around the screen. The presence of long lists of unbound issues from the serials check-in module, and bound volumes from item records, take up large portions of screen space, often requiring the user to scroll endlessly to reach the description for the next title. Summary holdings statements often appear rambling, separated by too many gaps and varying from title to title, and can be difficult to visually parse. Some of the data presented, such as the dates of publication statement from MARC tag 362, are confusing to users who interpret them as local holdings rather than summaries of publishing history. Citation notes, displayed from the 510, are misleading unless catalogers are constantly verifying that the journal truly is still indexed in that source. Maintenance of serial records is an ongoing commitment that most libraries probably cannot afford to make, though some have tried with e-journals. In many operations, a serial record is captured just once and remains in its captured state until it ceases, changes title, or is cancelled. Some means of linking this display with a common, well-maintained database, such as that of CONSER, would be useful. A number of libraries do this by subscribing to notification services from OCLC, which supply replacement records as the master file record changes. Ideally, this should solve typical serial problems, such as the library missing title changes and cessations, as well as changes to frequency, publisher, editors, and indexing sources. Of course this method must rely on the hope that at least one library will notice the change to a title's status and make the appropriate change to the master database record.

E-JOURNALS IN THE OPAC

E-journals have led to new challenges in the OPAC. Early e-journals, like those in JSTOR or Project Muse, were often reproductions of the print, much like the microform versions before them. As many libraries had already added microform holdings to their print serial records, they continued this practice for e-journals (the so-called MulVer model or single record approach). As publishers adapted to the electronic medium with features not supported by print, with additional content not found in the print edition, and even with title variations on e-versions, some catalogers began to turn away from this model, creating separate records describing first CD-ROM and then Web serials. While OCLC came out early in favor of separate records for all electronic versions, CONSER recognized that there were often legitimate reasons for combining versions on a single record, bestowing its blessing on both models in the 1994 edition of the *CONSER Editing Guide.* Serial records distributed by the U.S. Government Printing Office (GPO) follow both models, though largely combining formats on a single record. Another source of ambiguity in catalog treatment is the distribution of package records for batch loading by publishers and aggregators of electronic publications. Early on, these agencies sometimes repurposed readily available print records for serials, simply adding an 856 to link to their electronic version, much as e-book publishers had done. Today, some serials are ceasing publication in print entirely, continuing only electronically and posing a dilemma for catalogers holding out for combining print and electronic holdings on a single record. The result of all of these circumstances is that there is now a mix of practices in the library community, often within the same institution and catalog.

The muddle of varying e-journal practices on the publishing and cataloging sides is naturally mirrored in the way competing ILS vendors have handled e-resources and in the choices libraries have made locally to implement offered vendor solutions. While there were certainly hopes, as well as some vendor promises, that the ILS itself could handle all of the library's inventory control and resource discovery

needs for e-journals, many libraries have turned to ERMS applications, either as extensions of their ILS data or in standalone products
designed to manage the complex and unique issues accompanying
e-resources, such as licensing, link resolution, and holdings availability. These e-centric tools are increasingly forming the basis for
e-journal finding aids, using a combination of data extracted from the
catalog and data supplied by third-party license and subscription management services. Format-specific finding aids spun off from the catalog have existed for some time in the form of video and sound
recording catalogs, fiction lists, and local or union periodical listings.
Obviously, nonbook needles have been lost in the catalog haystack for
some time! Highlighting continuing resources, particularly e-journals
and databases, through interfaces outside of the catalog makes sense
in the context of timely public service. Since the financial meter is
ticking from the moment a license is signed, libraries want to make
these resources accessible to users right at that moment and to make
them inaccessible as soon as the license expires. E-resources represent a sizable investment in something that is often licensed rather than
owned; moreover, the access point of an e-resource is exact and frequently changing and cannot be identified or connected to by a patron
unless authenticated by a library's Web-based services. These are not
attributes that are easily managed through the catalog, yet in spite of
this most libraries continue to provide access to e-journals through
the catalog as well as through separate tools. Why is this so?

Probably the main reason why libraries continue to maintain point-
ers to e-journals in their catalogs has to do with the fact that many A-Z
lists and e-journal title search tools do not include print journal holdings, especially for ceased or cancelled titles. This means that the user
who comes to a journal title through such a list will be unaware of any
earlier volumes available in print only. These lists also frequently ex-
clude electronic government publications and other "free" resources,
which must still be searched in the catalog. Also, for many libraries,
the ILS continues to serve as the audit trail for financial transactions
concerning e-journals. Accordingly, there is a need to have at least a
brief record for these titles in order to track payment using the ILS.
Finally, e-journals remain in the catalog because the ILS is often the
source for the bibliographic data that is being displayed through the
A-Z list, E-journal finder, or other e-resource-specific finding aid.

In some cases, these external tools are refreshed periodically using data maintained in the catalog, while in others SQL queries or catalog searches are used against the ILS to provide more complete descriptive data than may be provided for in the serial-specific tool.

A BRIEF SURVEY OF ACCESS TO JOURNALS AND E-JOURNALS IN 30 ARL OPACS

As we have already seen, there are still valid reasons to keep serials, including e-journals, in the ILS. For print titles, government documents, and non-fee-based e-resources, this may be the only means offered the public for the search and retrieval of those titles held by the library. Also, there is the very real need to provide inventory control for print serials, including ordering, claims, binding, and current circulation status, attributes relatively well-handled in the ILS and for which there is considerable historic data. So, how are our leading academic libraries serving up periodicals through this interface?

In early October 2006, I surveyed the OPACs of the top thirty academic research libraries, as determined by the last ranking reported in the *ARL Membership Criteria Index* (the report is available at http://www.arl.org/stats/index/index05.pdf) for 2004/2005, along with the OPAC of Stanford University, which withdrew from ARL membership in January 2004 while ranked near the top.[6] All of these catalogs are based on an underlying ILS from one of six different vendors: Endeavor (ten libraries), Sirsi/Dynix (ten), ExLibris (five), Innovative Interfaces (five), and Geac (one). In almost all cases there has been at least some customization of the catalog interface to meet local periodical identification needs, though even in these cases, what is presented is determined largely by what goes into the catalog in the first place.

As Table 9.1 shows, all catalogs in the study supported both keyword and browse searches, and all included e-journals as well as print serials. All libraries supported separate e-journal-specific finding aids as well, either A-Z browse lists or title search tools, though these tools were not always accessible via links from the catalog itself. Often it was necessary to go back to the library's home page to find e-journal finding aids. A surprising 71 percent of catalogs provided links to e-journals from records describing print, while an overlapping

TABLE 9.1. Characteristics of the Surveyed Libraries

Feature	N	% of Total
A-Z list or searchable e-journal tool linked to from library Web site	31	100.00
E-journals in OPAC	31	100.00
Summary holdings for print	29	93.55
Filter by serial/periodical/journal	29	93.55
ISSN search	28	90.32
Periodical title search	27	87.10
Full-text/aggregated titles linked to resource	25	80.65
Summary holdings for e-journal	23	74.19
E-journal links from print record	22	70.97
Item (volume) records accessible for print	20	64.52
Separate E-only records	19	61.29
510 (Citation notes) display	19	61.29
E-journal provider/aggregator/vendor searchable	13	41.94
A-Z list or searchable e-journal tool linked to from OPAC pages	9	29.03

61 percent had separate electronic-only records. In three cases, there were separate records for each individual provider (JSTOR, Academic Search Premier, MasterFILE Premier, LexisNexis Academic, etc.) of an e-journal, including both reproduction and full-text versions. Two MulVer (or single record approach) libraries in the survey created separate "(Online)" entries in their title indexes by simply adding MARC 740s (added title entry) to the print record with the print title plus qualifier. Two others provided the same entries using the 776 (additional format) field.

One surprising finding is that libraries with records describing electronic-only versions often added URLs for those versions to print records as well, though five catalogs with separate print and electronic records distinguished between reproduction e-journals, such as those from JSTOR, and aggregator/full-text e-journals. In these cases, only the reproduction URLs appeared on print records, while the e-only records had both reproduction and full-text URLs. Since 74 percent of catalogs were attempting to maintain holdings for e-journals, including holdings from full-text databases, this implies a lot of duplicated effort. Plus, in some cases, holdings are also being maintained outside of the catalog through SFX, Serials Solutions, or TDNet.

The status of microform serials was not particularly enhanced by this study. While some libraries combined print, microform, and electronic formats on a single record, none of the libraries which cataloged each format separately provided online links from microform records, even when they did provide these links from print records. This may be for reasons of economy or it may reflect the libraries' perception of microform as an archival medium and patron choice of last resort.

The University of British Columbia's catalog was unique within the group of thirty-one as the only institution with no 856 links to any electronic versions. If one chooses the prominently-displayed "Journal/ Ejournal title search" option from the main catalog page, an intermediary screen is displayed where the user can choose to view the print version in the catalog or the electronic in SFX. If the print version has an electronic counterpart, the SFX link will appear at the bottom of the bibliographic display labeled "UBC eLink." Among serials, only those published by government agencies contain actual URLs in this catalog.

Some of the more noticeable features in some of the OPACs tested were the large buttons and other devices denoting the presence of an electronic version. These were not restricted to a single ILS vendor's products. A super-sized letter "e" denoted an e-version at both an Innopac site and another served by ExLibris' Aleph. A similar large, round "networked resource" button adorned the index of an Endeavor site. Texas A&M displays a rather large "Full Text at TAMU" button at the top of the bibliographic display for a title that SFX has determined as having an electronic counterpart which the university has licensed. A less visible device denoting the presence of an electronic version comes from the University of Virginia's brief display, which simply states "URL" after the label "A look inside."

Finally, as an experiment in title searching, all thirty-one catalogs were searched for the elusive newspaper, the *New York Times*. To be consistent, each OPAC was searched by title keyword and, in many cases, by title browse, and the equivalent of periodical title. Obviously, when one simply enters words in a keyword title search, the OPAC should not know that you are looking for a periodical, yet in one Sirsi catalog, the expected newspaper record was the first entry to appear in a hitlist, for both keyword and browse searches! In two Endeavor catalogs it came up either first or second in keyword searches. The only Geac catalog ranked the *Times* third, while in all of the other catalogs, it was nearly impossible to find the print newspaper record. In some displays, one had to know the date of first publication (1851 in some, 1857 in others) to find the entry; in others, one had to scan through hundreds of alphabetically sorted responses to retrieve it.

CONCLUSION

The survey shows that serials, both print and electronic, are still being entered and maintained in libraries' catalogs, in spite of what are viewed as obvious limitations to that tool in terms of search, retrieval, display, and maintenance. None of the top thirty ARL libraries, has been willing to make a clean break with the OPAC where serials are concerned. This may be a case of libraries hedging their bets on which technology will ultimately prove itself most useful given the relative youth of e-journal identification tools and the trust most place in the ILS, at least for data storage. The matter of data storage should not be treated lightly here, as the most enduring features of the MARC standard have been its granularity, standard coding schema, and perhaps most importantly, its portability when the next generation of products emerges.

Nonetheless, there are trends in the marketplace that are hastening the demise of the catalog, especially as a tool for providing access to e-journals. Library economics is certainly the primary driving force here, with maintenance of description, access, and holdings in two or more places not likely to survive for long in today's economic climate. It is likely that libraries are just unwilling to face this fact yet, but just as likely that they will have to soon, as new services, both internal and external to the library, compete for the money that we are spending on cataloging and catalogs. The recent Calhoun report paints a bleak future for the catalog as it competes with the open Web for the attention of the information-seeking public and fails to achieve the economies necessary to survive in a shrinking market for its services.[7] Calhoun argues that libraries need to be strategizing now on what the catalog is good at, how we can make it better and cheaper to maintain, and what we might have to give up if we expect to retain it at all. With more periodical backfiles being added to existing e-journal offerings, patrons voting for e-journals with their feet, and physical space at a premium, libraries are cancelling print subscriptions, perhaps pushing us closer to a time when almost all serial literature will be online and the tools for its retrieval will be outside of the catalog and perhaps outside of the library's control.

NOTES

1. Andrew Pace, "Catalogs for the Future," Presented at the 2006 Computers in Libraries Conference, 2005, http://www.lib.ncsu.edu/endeca/presentations/200603-endeca-pace2.ppt (accessed December 7, 2006).

2. Roy Tennant, "MARC Must Die," *Library Journal,* October 15, 2002, http://www.libraryjournal.com/article/CA250046.html (accessed December 7, 2006).

3. Karen Coyle, "Is MARC Dead?" Presented at an ALA Annual Conference 2000 panel, http://www.kcoyle.net/marcdead.html (accessed December 7, 2006); Kristin Antleman, Emily Lynema, and Andrew K. Pace, "Toward a Twenty-first Century Library Catalog," *Information Technology and Libraries* 25, no. 3 (2006): 128-139.

4. Ibid., 128.

5. Maria Collins et al., "Magnifying the ILS with Endeca," *Serials Librarian* (2006, forthcoming).

6. "University Withdraws from Library Association," *Stanford Report,* January 30, 2004, http://news-service.stanford.edu/news/2004/february4/arl-24.html (accessed December 7, 2006).

7. Karen Calhoun, *The Changing Nature of the Catalog and its Integration with Other Discovery Tools,* a report prepared for the Library of Congress (2006), http://dspace.library.cornell.edu/bitstream/1813/2670/1/LC+64+report+draft2b.pdf.

Chapter 10

ERM Systems: Background, Selection, and Implementation

Maria D. D. Collins

INTRODUCTION

Electronic Resource Management (ERM) systems are inundating the library marketplace. Both Integrated Library System (ILS) vendors and Public Access Management Services (PAMS) offer commercial options for their library customers. In addition, homegrown and open source solutions are also part of the ERM landscape. So, why do libraries need an ERM system?

E-resources have dramatically changed academic library workflows. The volume of e-resource materials collected in libraries has reached a critical mass that prohibits traditional title-by-title management. The single-title subscription is no longer the only means for acquiring access to journal content. In today's library, purchasing collections of titles in aggregated full-text databases or publisher packages is common practice. Therefore, the primary unit of acquired content is evolving along with the format of the collection. In this day and age, it is not unusual for an academic collection to be over 50 percent electronic. The changing complexion of academic research collections requires definite changes in the workflows used to effectively manage these materials. E-resources tend to be surrounded by invisible tasks that are difficult to track. With an e-journal, there is often no physical reminder to perform tasks such as license negotiations and authentication. It is up to the library to create a system for managing these tasks. In addition, effective follow-up requires some kind of mechanism that is not memory bound. The e-resource process is nonlinear in nature, time-consuming, and multilayered. Even the brightest professional would have trouble keeping all the necessary information or procedures in mind to handle the complex life cycle of e-resources without some kind of tool to aid this process.

ERM systems are being designed to fill this void in management. Libraries want to systematically integrate the many processes and tools used to manage e-resources. Optimally, an ERM system should provide one point of maintenance or serve as a single container of information to perform a variety of functions including public display, license management, collection evaluation, reporting and statistical analysis. These are often functions that traditional tools such as the ILS are not designed to carry out. There are other potential advantages to ERM systems that make them attractive to libraries. These systems may facilitate workflow processes by pushing various tasks or sending out reminders using ticklers or queues. They should streamline workflows by eliminating duplication of effort through the maintenance of multiple spreadsheets and databases many libraries previously used to manage e-resources. Furthermore, in addition to keeping track of the license status, many systems can display usage rights for patrons and help with the resolution of contract breaches through the use of logs. Depending on the data model of the system and its integration with the ILS, these systems may also assist the cataloging process through their ability to create brief records. Of course, depending on the ERM system and a library's current systems infrastructure, many of these perks may still be under development.

Given the potential value these tools provide in managing academic collections, it is imperative for libraries to understand and evaluate ERM systems in respect to their priorities. To facilitate this understanding, this chapter will begin by reviewing the background and history behind ERM development and highlighting current initiatives. Next, the chapter will discuss the selection of an ERM and provide advice for implementation. This information will hopefully provide a realistic perspective concerning these tools and reveal how ERM systems fit within a library's existing framework, including workflows and personnel. Simply put, for all the advantages these tools may provide, they are just part of the equation for effective e-resource management.

BACKGROUND AND HISTORY

In order to tackle the issues associated with e-resource management, many libraries began developing library-based or homegrown ERM systems in the late 1990s and early 2000s. These systems were designed to provide functionalities missing from existing library tools and services such as robust reporting, detailed acquisitions information for e-resources, and licensing description. The report of the Digital Libraries Federation's Electronic Resource Management Initiative (DLF ERMI) notes that over twenty homegrown systems were established or in development between 2001 and 2003.[1] Several resources—such as the DLF ERMI report and Geller's *Library Technology Report,* titled *ERM: Staffing, Services and Systems*—discuss local initiatives and their impact on the development of the DLF ERMI.[2] Additional explanation of key features of several of these homegrown systems, including Penn State's ERLIC, MIT's VERA, and UCLA's ERDb, are discussed by Tim Jewell in his recorded NASIG presentation on e-resource management in 2004.[3] These local systems provided a variety of functions unique to the individual library that developed the tool. Quickly, libraries realized that they were dealing with many of the same issues and concerns specific to e-resource management, and collaborative efforts evolved.

In 2001, Adam Chandler from Cornell University and Tim Jewell from the University of Washington created a Web Hub (http://www.library.cornell.edu/elicensestudy/webhubarchive.html) to "exchange information about local systems and foster communication among interested librarians."[4] Other synergies evolved from this effort including Tim Jewell's publication "Selection and Presentation of Commercially Available Electronic Resources" (http://www.clir.org/pubs/reports/pub99/pub99.pdf), which provides an analysis of local systems' functions and data elements.[5] In addition, the realization that standards were needed to further the development of these systems ultimately led to the creation of the DLF ERMI in 2002. The primary purposes of this initiative were to formalize this effort, provide a description of the functional requirements desired in an ERM, identify data elements and common definitions for consistency, provide potential XML schemas, and identify and support data standards.[6] The achievement of these goals would facilitate and guide the development of existing and future ERM systems. Facilitating the development process, minimizing vendor costs, furthering capabilities for data exchange, and future system migrations were additional motivations behind the DLF ERMI.[7] Essentially, when published in 2004, the DLF ERMI report provided a blueprint for the preferred features and system design of an ERM. This standard has since been adopted by most commercial vendors providing an ERM system.

Phase II of the Electronic Resource Management Initiative (ERMI 2) is well underway and continues to support the development and utilization of ERM systems. The DLF Electronic Resource Management, Phase II Web site (http://www.diglib.org/standards/dlf-erm05.htm) explains that the "second phase of this project capitalizes on and extends the visibility and success of ERMI with a particular focus on data standards, issues related to license expression and usage data."[8]

One example of the continued focus on data standards is EDItEUR's (http://www.editeur.org) adoption of ERMI's work on data elements to further the development of its license expression initiative. EDItEUR, which is an international group focused on standards for electronic commerce, is working to establish a "publishing industry license messaging standard within the ONIX family of transmission standards."[9] This standard, called ONIX for Licensing Terms, aims to "express [license] terms in a standard XML format, link them to digital resources, [and] communicate them to users."[10] Currently, this group is investigating the use of the ERMI data elements to facilitate this process. A proof-of-concept project that began at the end of 2004 is in development to examine the specifics of how this transfer can occur and create a prototype for communicating the terms of use in a license agreement.[11] Additional description of the development and requirements of this data exchange project can be found on the EDItEUR Web site.

ERMI 2 has also established a training initiative that focuses on mapping license terms to ERM systems. These sessions have been held at a variety of venues including American Library Association (ALA) conferences and the North American Serials Interest Group

(NASIG) conference. Their initial focus has been on analyzing usage rights in license agreements and establishing consistent methods of interpreting these terms to enter into an ERM system.

In respect to the third focus of ERMI 2 mentioned in previous text, usage statistics, several members of ERMI are involved with the transfer of vendor data into ERM systems through active participation with Standardized Usage Statistics Harvesting Initiative (SUSHI) (http://www.niso.org/committees/SUSHI/SUSHI_comm.html).[12] This initiative seeks to establish a "standard data container for moving Project COUNTER usage statistics into a digital repository."[13] An ERM system can serve as one example of an appropriate repository to receive this data. Ultimately, anyone who has recently implemented an ERM system recognizes the need for further development in data standards and interchange; Phase II efforts are extremely important in furthering these capabilities in ERM systems.

CHOOSING THE RIGHT ERM SYSTEM

The Players

Understanding the development behind ERM systems is one step toward making an appropriate selection from the tools available to meet a library's ERM needs. The selection process is perhaps the most difficult aspect of implementation for many librarians and requires the knowledge of potential systems to purchase, a familiarity with the library's current ERM tools, and an understanding of the library's priorities for e-resource management. This section will explore these criteria to better illuminate the selection process.

First, who are the players offering ERM systems and what systems are they offering? ILS vendors first entered the marketplace in 2004 with the general release of Innovative's ERM System.[14] Other ILS vendors soon focused their efforts on developing competing systems. As of the end of 2006, ILS vendors offering ERM modules included the following:

- Endeavor's Meridian—(Supported until the end of 2008)
- Ex Libris's Verde
- Innovative's ERM

- Sirsi/Dynix's ERM Solution through Serials Solutions
- VTLS's Verify

A few partnerships and mergers are worthy of note from this past year. In April 2006, SirsiDynix announced its partnership with Serials Solutions.[15] Serials Solutions ERMS will be integrated into both the Horizon and Unicorn products from SirsiDynix.[16] Also, in November 2006, the parent company of Ex Libris, Francisco Partners, finalized its acquisition of Endeavor.[17] Endeavor's ERM system Meridian will continue to be supported until the end of 2008. Ex Libris will support Meridian customers migration to Verde if they so choose.[18]

Even though most of the companies noted in previous text are able to sell their products as stand-alone systems, all offer or will offer their ERM system as an integrated component of their ILS. If a library selects an ERM system that is integrated with its ILS, there are several advantages, chief among these being interoperability. Indeed, existing data in the ILS should be easily associated with data from the ERM system. Single points of data entry should support functions of multiple modules; for instance, payment information noted in the acquisitions module should be easily extracted and displayed in the ERMS or paired with ERM data for reporting purposes. In addition, often with integrated systems, the vendor will enhance other modules to support the functionality of its ERM module. These systems will also provide users with a consistent interface. Collins notes that "providing library staff with a consistent interface for all e-journal functions should minimize training and enhance usability."[19]

Of course, if a library uses the same vendor for all of its systems, ILS and ERM systems included, then the library is at the mercy of that vendor for timely updates and improvements. As a result, a library could end up with a system that is under-supported. This further emphasizes the importance of knowing each vendor's track record on development including the known strengths of a vendor's systems. If a library chooses a stand-alone system or differing vendors for its ERM system and ILS, there could be problems in respect to integration. Even if the potential for integration exists, local programming expertise may be required to connect various systems. Perhaps the greatest disadvantage of using an ILS vendor to provide an ERM system is the lack of a knowledgebase or title management database as supplied by vendors offering link resolvers or A-to-Z listing services. Title and subscription details would still be required to populate an ERM. As the ERM systems of ILS vendors mature, integration with third party link resolvers and A-to-Z services to obtain access to a knowledge-base will most likely become less of an issue. In fact, partnerships between ILS vendors and PAMS such as the SirsiDynix/Serials Solutions partnership may change this dynamic, since, for instance, SirsiDynix customers will have access to an integrated ERMS solution with the full benefits of a knowledgebase, when fully developed. When ILS vendors were asked if their ERM systems currently interoperate with outside data sources, most of these companies stated that this was possible.[20] However, internal systems resources for scripting may still be necessary to make these varying systems talk.

Unlike ILS vendors, systems offered by third parties—such as PAMS or nonprofit organizations like CARL's Gold Rush—do provide a knowledgebase as part of their suite of services. Currently, the non-ILS, third-party companies or organizations offering an ERM system include the following:

- Serials Solutions ERMS
- TDNet TeRM
- CARL's Gold Rush

These ERM providers have built their credibility on their ability to link and manage a given library's subscriptions, including resources available via aggregated databases, through OpenURL link resolvers or A-to-Z listing services. Therefore, some library customers may feel that they can better trust their data in the hands of one of these companies. Collins further explains this scenario stating that "users may feel that third-party systems are the better option for supplying their ERM system because they are more experienced with the complexities of e-journal data."[21] An additional advantage for libraries who are already subscribing to multiple ERM tools from the same third party, such as a link resolver or A-to-Z listing service, is the ease of sharing the knowledgebase across multiple tools.

There are, of course, disadvantages to using this type of vendor. Similar to the stand-alone ILS systems described previously, integration with a library's ILS may be questionable for these systems. There is also the issue of integration with other third-party ERM tools (i.e., link resolvers, metasearch tools, A-to-Z lists). During the selection process, a library should carefully examine and test whether or not these ERM systems can coexist with the library's existing systems infrastructure. Do these ERM tools allow the export and import of data? If so, which file formats are supported (MARC, delimited file formats, XML-based formats)? Is interaction with a library's ILS possible, and if so, is custom mapping required? Most of the ERM vendors have stated that their systems have these capabilities or that they are in development.[22] Careful discussions with customers already implementing these systems may be necessary to determine if the systems' level of functionality will meet a library's needs.

Subscription agents are another category of ERM services that may or may not fit into a library's e-resource management infrastructure. There are three agents who either support the e-resource management process or provide ERM systems: EBSCO's Electronic Journal Services (EJS), Swets SwetsWise, and Harrassowitz HERMIS. The use of these services is tied to a library's utilization of these companies' subscription services. Often a library can choose to access basic management services as a current customer or elect to pay an additional fee for advanced services. These agents are extending their services beyond print and are able to offer a knowledgebase of subscription details, but they may not be able to comprehensively support all of a library's e-subscriptions or e-access if a library uses another vendor's A-to-Z or link-resolving tools or if a library uses more than one subscription agent. Essentially, the ERM functionality provided by these companies builds on subscription services. Duranceau describes these agent services in her June 2005 review as "enhanced subscription support for e-journals, along with some basic reporting or listing/access tools for e-resources."[23] These services can be utilized in a number of ways: they can provide supplemental data to a separate ERM system, support public display functionality, or even facilitate the registration/authentication process. Obviously, their use is not exclusive of other tools.

Of course, even with this expanding field of commercial options, a library may choose to either develop its own ERM system or adapt an open access solution. Currently, there is one open access ERM system noted in the DLF ERMI report, Johns Hopkins' Hermes system.

Jewel explains that this system is worth noting due to its "careful analysis of staff roles, workflows, and associated functional requirements."[24] The choice to grow your own ERM or support an open access option is reliant on sustained systems resources by the library making this selection. Successful implementation of either option will directly correlate to continued development by the library and the level and sophistication of programming resources available. If, indeed, a library has existing systems and programming resources, these options are advantageous in that the "systems will be tailored exactly to local needs."[25] In a recent article focused on choosing to grow your own ERM, Stephen Meyer suggests just this advantage in respect to North Carolina State University (NCSU) Libraries' decision to continue to support their in-house system E-matrix.

The discussion of the systems in previous text excludes descriptions of specific features or functional elements. This information has been detailed in several helpful articles spanning the last few years, including the following:

- Duranceau's "Electronic Resource Management Systems from ILS Vendors" published in *Against the Grain,* September 2004;
- Collin's "Electronic Resource Management Systems: Understanding the Players and How to Make the Right Choice for your Library" published in *Serials Review,* June 2005;
- Duranceau's "Electronic Resource Management Systems Part II: Offerings from Serial Vendors and Serial Data Vendors" published in *Against the Grain,* June 2005;
- Meyer's "How Many ERM Systems are out There?" published in *Computers in Libraries,* November/December 2005; and
- Geller's *Library Technology Report* titled "ERM: Staffing, Services, and Systems" published in March/April 2006.[26]

These resources relate descriptions of ERM systems available on the market including offerings from ILS vendors, PAMS, subscription agents, and nonprofit organizations. The methods used by these authors for obtaining ERM system data range from informal interviews to surveys. For a quick reference of the ERM systems available, see Table 10.1.

TABLE 10.1. ERM Systems Available As of December 2006

ERM	Company	Type of Company	Integrated or Stand-Alone[a]
Meridian	Endeavor-merged with Ex Libris	ILS	Integrated with ILS or stand-alone. No longer supported after 2008
Verde	Ex Libris	ILS	Integrated with ILS or stand-alone
ERM	Innovative Interfaces, Inc.	ILS	Integrated with ILS or stand-alone
ERM interface with Serials Solutions	SirsiDynix	ILS/PAMS	Integrated with both Horizon and Unicorn products
Verify	VTLS	ILS	Integrated with ILS or stand-alone
ERMS	Serials Solutions	PAMS	Stand-alone. Will be integrated through SirsiDynix's ILS systems
TeRMS	TDNet	PAMS	Stand-alone
Gold Rush	CARL	Nonprofit company	Stand-alone
EbscoHost's Electronic Journal Service	EBSCO	Subscription Agent	Compliments other ERM tools
SwetsWise	Swets	Subscription Agent	Integrated with subscription services
Hermis	Harrassowitz	Subscription Agent	Integrated with subscription services
Hermes	Johns Hopkins	Open Source	Stand-alone

[a]Integrated implies that system is integrally tied to a company's suite of products. Stand-alone implies that system is available for purchase as a solitary product.

UNDERSTANDING YOUR LIBRARY'S ERM NEEDS AND LIMITATIONS

An academic library's e-resource management needs are tied to the nature and mission of its parent institution, which in turn determines the size of the library's collection and intended areas of growth. As libraries develop their technological infrastructures, a careful balance must be obtained between the cost of these advancements and benefits to users. For example, if a small liberal arts school only has an e-journal collection of a few thousand titles, then the ERM tools used to manage these resources can be minimal. In contrast, a large academic university with an e-journal collection of 25,000 and a student enrollment of 20,000 has a greater need for ERM tools. Therefore, size and type of library and institution cannot be ignored in determining a given library's ERM needs. In addition, the financial resources available to the library are also an important factor in the selection of a library's suite of tools.

Factors such as size, type, and financial resources provide a basic framework for decision making. Another important consideration that also frames an ERMS decision is the current systems infrastructure already in place. Existing tools and resources used for management—including metasearch tools, link resolvers, ILS, title-tracking services, etc.—should be identified and assessed. Interoperability with these existing systems should be determined to maximize the greatest benefit from adding an ERM system. In addition, programming and systems expertise should be identified and allocated to allow for a successful implementation of the ERM system; otherwise, the library may end up wasting funds on a tool with extra bells and whistles that also requires extensive local customization. This same consideration may also determine a library's ability to host an ERM system on its own servers. Several vendors offer to host the system for the library if this is desired.

The workflows established in a given library also provide insights into priorities for e-resource management from the frontline perspective of staff working with these resources each day. Established procedures indicate prior areas of focus, problem areas that have been addressed through work-arounds, and existing limitations of the library's resources. All these factors serve as an excellent starting point in analyzing a library's needs for an ERM system. What problems does your library hope to resolve; what outcomes does your library hope to accomplish? A basic checklist of desired functionality follows that can also help in this analysis process. Librarians can check this list against both their existing processes and desired outcomes to help determine which ERM systems are best able to meet their library's unique needs. Note that this checklist is not an exhaustive presentation of features available in ERM systems, but simply reflects potential workflow areas where libraries have concentrated their efforts to handle the e-resource life cycle. Another useful reference for analyzing desired functionality for an ERM system is Appendix A. Functional Requirements for Electronic Resource Management of the DLF ERMI report.[27]

ERM Checklist

- License Management
  - —Display of selected license terms to the public
  - —Container for license description, ease of access, and analysis
- Workflow Management
  - —Queues for distributing e-resources tasks
  - —Ticklers/reminders to push tasks through life cycle
- Reporting
  - —On-the-fly reporting for custom reports
  - —Canned reports for core reporting functions
- Usage Statistics (with import and export functionality)
- Collection evaluation
  - —Selection
  - —Facilitating selection process
  - —Tracking selection decisions
  - —Collection analysis and statistics
  - —Management of trials
- Acquisitions features
  - —Cost
  - —Purchase orders
- Cataloging functionality
  - —Create brief records
  - —Facilitate MARC record management
- Public display features
  - —Management of library subject pages
  - —Management of library A to Z list
  - —Management of library database list
- Authentication/Registration
- Security/User Profiles for Access
- Administrative Tasks
  - —Problem logs
  - —Technical, administrative contacts

Obviously, given the variety of tasks noted in this list, there are a large number of considerations a library must reflect upon when choosing the most appropriate ERM system to meet its needs. As Ownes states, "conducting an internal and external needs assessment is one of the best things librarians can do when considering an ERMS purchase."[28] The selected system should complement existing systems and provide solutions for the most pressing issues particular to that given library in respect to e-resource management. Too often, a library will make the mistake of selecting tools that appear to be the "best" system on the market rather than the best fit for its institution, which may result in a system that is not carefully integrated into the library's existing systems infrastructure. Libraries have to take the initiative to ensure that they are driving these management initiatives through careful planning and selection rather than purchasing a system and discovering its functionality and limitations after the fact.

IMPLEMENTING AN ERM SYSTEM

Planning—Staffing, Communication, and Workflows

Once a library has a clear understanding of its desired purposes for an ERM system and has purchased the most appropriate tool, the extensive implementation process begins. This process can be extremely time-consuming, depending on the number of desired features the library wishes to utilize. Careful planning of human resources, communication strategies, and existing workflows is necessary for effective implementation.

In order to fully utilize an ERM system, extensive manpower is usually required. Therefore, staffing is one obvious concern during the planning process for implementation. As discussed previously, the systems staff available to maintain and customize an ERM system are integral to the type of system a library decides to purchase. In addition, a library will need staff to implement, populate, and work with the system. Kasprowski further emphasizes the number of staff needed for ERM processes, noting that "any staffing issues in the online age should cause a shift in responsibilities, not loss of work, as e-resource management, in general, and ERM system implementation, in particular, are work intensive processes."[29] Therefore, during the initial stages of implementation, a library should consider who should be involved in these processes and how personnel will be structured within the library, whether individuals assigned to ERM tasks should be integrated within existing departments or whether a separate unit needs to be created. No matter which organizational approach a library decides to take, staff training and continuing education will be a definite priority to address. This is especially true since e-resource processes affect so many different areas of the library. Staff assigned to these tasks may have varying backgrounds and experience with e-resource management tools.

Communication across departments is also key to a successful implementation. Knowledge concerning e-resource processes cannot be contained within one or two positions or in one department if librarians or staff from outside departments are to make educated decisions across various aspects of the e-resource life cycle. For example, if an acquisitions department is the negotiator of a library's license agreements, then any Interlibrary Loan (ILL) restrictions need to be communicated to the individuals or departments handling ILL. ERM systems can facilitate this communication process if appropriate communication channels are discussed and created during implementation.

Compartmentalized information can also be problematic within the departments responsible for most of the e-resource management activities. Often requiring specialized knowledge or unique handling, e-resource processes may have initially been assigned to key individuals rather than being distributed across multiple positions or departments. As the number of e-resources added to a library's collection reaches a critical mass, e-resource staff will need to break down their specialized information silos and communicate more broadly to better integrate these processes across additional positions. Establishing cross communication strategies when implementing an ERM system will help ensure that e-resource tasks are not person-specific. Routine procedures can then evolve to handle ERM tasks. Given the invisible nature of many e-resource tasks, distributing these responsibilities in addition to using an ERM system to create automatic reminders of various processes is integral to their successful management. The system created for handling e-resource workflows and processes should not be memory-bound; instead, it should allow for communications independent of individual staff.

These communication strategies are closely aligned and instrumental to creating successful workflow strategies desired for the implementation of an ERM system. The workflows established through an ERM system should incorporate existing procedures and create a natural order for the library departments utilizing the system. It is important that the system grows and adjusts to the chaos of ERM tasks rather than seeking to create an imposed structure for these tasks. This will allow the system to evolve as e-resource processes change. In order to plan for the workflow changes that will occur as a library implements an ERM system, careful evaluation of existing workflows is necessary. Each process directly or indirectly affiliated with e-resource management should be questioned and its value calculated and balanced against the new efficiencies a library hopes to gain through the implementation of an ERM system. This reflection will reveal procedures to discard or retain. This evaluation process will also free up resources to incorporate the changes brought about by the new system. Integrating an ERM system into existing workflows should also include plans to integrate with current ERM tools. Essentially, the successful implementation of an ERM system rests on the staff's ability to fully incorporate this tool and not treat the system as an "add-on" resource outside the mainstream of departmental processes.[30]

Additional workflow strategies to consider during the planning stages of implementation center on establishing data standards for local practices as well as for consistent use of such national standards as the ERMI guidelines. If the fields of a library's selected ERM system do not exactly replicate national standards, consistent use of data will ensure the library's ability to map this data to recognized standards in the future. Furthermore, consistent data entry practices will help to facilitate any future system migrations. To ensure this kind of consistency, librarians need to be aware of new developments and current standards; local standards should be clearly documented and communicated, and a strong understanding of data and data sources must exist. Careful attention to these details during the implementation process will allow data exchange opportunities later on.

Mapping license agreements is another area requiring significant attention to detail and planning before and after implementation. The ability to describe and analyze license agreements may be one of the primary reasons your library purchased an ERM, but this one task is extremely time-consuming and complex to accomplish even if using the ERM to evaluate license terms afterward will be a quick process. Many librarians focusing on this eventual outcome of the license data entry process may completely underestimate the time it takes for any institution to carefully determine how to consistently read and record variations in language for a wide range of license terms. Phase 2 of the ERM Initiative addresses this very issue in its continuing education workshop series, which outlines ERMI's data dictionary for license terms and provides hands-on practice in the mapping process for analyzing usage rights language (e.g., ILL and reserves) typically found in license agreements.

Given these complications, it becomes extremely important for a library to plan how it will map license agreements. The following questions should be addressed before diving head first into this exhausting process:

- What elements are important to include for your library?
- What elements are repetitive across license agreements and provide little value or are inconsequential in describing?
- Who will be responsible for providing consistent interpretation of license language and meaning?
- What tools or resources are available to assist individuals in the mapping process?

Once library staff have answered these questions, they can begin developing a process for mapping license data. Creating a form or template for review and analysis is a helpful strategy for assisting this process. Librarians hosting the ERMI workshop do provide a useful form for assisting in the license review stage. Appendix 10.A is a sample of a form used to assist the mapping process at a large academic library (NCSU Libraries).

Advice from Vendors and Librarians

Probably one of the best strategies to gear up for implementation is to seek advice from the vendors creating these systems and librarians that have already implemented an ERM system. In a June 2005 *Electronic Journal Forum* column, Collins notes advice obtained from ERM vendors concerning steps to take before beginning implementation. The following are a few of the tips that were provided:

1. Ask data providers to support and use data standards
2. Work with the e-resource community to create industry-wide standards that will assist in the exchange of acquisition and e-resource data
3. Implement other e-journal management tools such as link re-solvers or A-to-Z tracking services
4. Scan or digitize license agreements
5. Determine the most important data elements to meet local needs
6. Investigate data sources and begin gathering data to load
7. Standardize data
8. Investigate details for migrating local ERM systems to a com-mercial ERM system
9. Ensure that your library can provide the appropriate systems en-vironment to support an ERM system.[31]

Further advice and discussion on ERMs were provided through an informal e-mail survey conducted on the SERIALST listserv in May 2006, which asked about the various stages of implementation, prob-lems encountered during this process, a library's primary purpose for obtaining an ERM system, the greatest strengths of various systems, future improvements librarians would like to see, and finally, advice for implementation. The responses to the first question revealed that the various stages for implementation basically mimic those used to implement any new system. One response outlined the stages of im-plementation in the following way:

1. Identify sources for ERM data
2. Identify and order matching ERM elements (ISSNs, titles, etc.)
3. Identify staff
4. Assess training needs
5. Draft workflows
6. Consult with other players (Acq, Cat, etc.)
7. Establish documentation procedures
8. Trial run of workflows
9. Review the process
10. Revise as necessary, repeat[32]

When asked what libraries were looking for in an ERM system, responses ranged from seeking a system that would integrate with a library's current ILS to looking "for ways to automate the ingestion of data" using standards such as ONIX for Serials and SUSHI.[33] Another answer to this question mentioned desired outcomes including tracking problems with vendors, renewal alerting, displaying license terms for patrons, and managing trial processes and history.[34] A similar question, focusing on the primary purpose for obtaining an ERM, generated some of the same responses noted in previous text with one exception. Indeed, one librarian noted that he was seeking a system that would both facilitate communication and workflows. He further stated that "Any ERMS that doesn't facilitate communication and workflows isn't worth the sticker price."[35]

Problems encountered during implementation also varied and seemed to reflect more broad issues with implementation rather than being specific to any particular ERM system. Problems noted included the failure to handle consortial aspects due to DLF ERMI specs not addressing this area, being overwhelmed by the volume of manual data entry required, having difficulty incorporating the tool and processes into everyday workflows and priorities, and misjudging the number of staff needed from different departments to assist in implementation.[36] Other concerns noted were the "lack of one-to-one mapping between source and destination ERMS,"[37] not prioritizing populating the system, and staff not understanding the value an ERMS would provide to their library.[38]

Alternatively, the strengths noted tended to be system-specific. Medeiros noted that the VTLS system "facilitates in a robust way the various workflows that revolve around e-resources."[39] A Serials Solutions user, Holmberg, mentioned that working with a company that is not an ILS was advantageous since PAMS have different priorities and can work easily with more libraries. She also notes the responsiveness of Serials Solutions and their user-friendly interface.[40] Matthews, who uses Innovative's ERM, notes the "flexibility of the system to load e-resources other than serials" as a plus for this system.[41]

Future improvements desired primarily focused on an ERM system's ability to import data such as usage statistics, license terms, or other kinds of digital or electronic information.[42] Other features these respondents wanted to see in future versions of their ERM systems included better integration with the ILS (in particular, the acquisitions module) and increased ability for customization.[43] All of these suggestions reflect a need for increased flexibility in an ERM system.

Finally, these survey responses did provide a few words of advice for those just beginning the implementation process. Of course, planning ahead in addition to understanding the library's ERM needs was stressed. Moreover, librarians should be realistic about the scope of their ERM needs and match their solutions and tools to the complexity of the ERM services that are required.[44] A couple of respondents noted the importance of "canvassing" the ERM community including questionnaires for vendors and surveys for users of the system.[45] Make sure that the ERM systems of interest have been fully investigated through contacts that know those systems well. This will help to facilitate the evaluation process. One last point about the implementation process emphasized the patience required to carry out such an undertaking. Given the scope of an ERM system, a concerted effort must take place to prioritize how this system will be incorporated into the staff's daily routines. This may mean adjusting current routines to accommodate the increased workload. In fact, Matthews notes that his library temporarily halted certain tasks to allow time for the implementation. He discusses the importance of focusing all efforts on the implementation process stating the following:

> Trying to implement such a complex service that required cooperation throughout an organization (the departments of which are in competition for resources—people, time, money) is best serviced if you hit the ground running. Create a big splash at the outset, so people can experience the positive impact of ERM on users, and then parlay that experience into leverage for resources.[46]

This last point of buy-in is an important one to ensure the success of these tools. An ERM system is not a tool that only requires initial resources upfront. Continued support and staffing are essential for developing the product and exploring additional uses and benefits.

CONCLUSION

ERM systems are only part of the equation for effective e-resource management. Personnel resources, other tools that make up the ERM infrastructure, in addition to the workflows established to carry out e-resource-related tasks are all part of the management process. Once an ERM system has been evaluated and carefully selected, its success hinges on the ability of library staff to incorporate this tool into their daily routines and integrate it with existing systems. The value an ERM system provides is severely limited if it functions as an "add-on" tool on the periphery of existing workflows. Furthermore, the selected system should facilitate and help libraries achieve desired goals, which should be predetermined through a needs-assessment process. The functionality of the system (or lack thereof) should not be the deciding factor driving departmental procedures and policies. Careful selection and planning will improve a library's chance of implementing an ERM system that can evolve with e-resource workflows and industry initiatives while at the same time addressing library priorities.

NOTES

1. Timothy D. Jewell et al., "Electronic Resource Management Report of the DLF Initiative," August 2004, http://www.diglib.org/pubs/dlfermi0408/ (accessed November 30, 2006).

2. Ibid.; Marilyn Geller, "ERM: Staffing, Services and Systems," *Library Technology Report* 42, no. 2 (March/April 2006): 15-16.

3. Timothy D. Jewell, presenter and Anne Mitchell, discussion recorder, "Electronic Resource Management: The Quest for Systems and Standards," *Serials Librarian* 48, no. 1/2 (2005): 140-144.

4. Ibid., 140.

5. Timothy D. Jewel, *Selection and Presentation of Commercially Available Electronic Resources* no. 99 (Washington, DC: Digital Library Federation and Council on Library and Information Resources, 2001), http://www.clir.org/pubs/reports/pub99/pub99.pdf (accessed December 1, 2006).

6. Jewell, "Electronic Resource Management Report."

7. Jewell and Mitchell, "Electronic Resource Management," 147.

8. Digital Library Federation, "DLF Electronic Resource Management Initiative, Phase II," July 12, 2006, http://www.diglig.org/standards/dlf-erm05.htm (accessed November 5, 2006).

9. EDItEUR, "EDItEUR," http://www.editeur.org/ (accessed November 5, 2006); Digital Library Federation, "DLF Electronic."

10. Ibid.

11. EDItEUR, "Report on ONIX for Licensing Terms Proof of Concept Project," http://www.editeur.org/ (accessed November 5, 2006).

12. Digital Library Federation, "DLF Electronic."

13. NISO, "NISO Standardized Usage Statistics Harvesting Initiative," http://www.niso.org/committees/SUSHI/SUSHI_comm.html (accessed November 5, 2006).

14. Maria Collins, "Electronic Resource Management Systems: Understanding the Players and How to Make the Right Choice for Your Library," *Serials Review* 31, no. 2 (June 2005): 128.

15. SirsiDynix, "SirsiDynix Partners with Serials Solutions for Integrated E-Resource Management and Discovery," http://www.sirsidynix.com/Newsevents/Releases/releases_2006.php#apr06 (accessed December 15, 2006).

16. Sharon Dyas Correia, reporter, "21st Conference: User Group: Sirsi Dynix," *NASIG Newsletter* 21, no. 3 (September 2006): http://nasignews.wordpress.com/2006/09/03/213-200609-21st-conference-user-groups-sirsi-dynix/ (accessed December 15, 2006).

17. Ex Libris, "Ex Libris and Francisco Partners Complete Acquisition of Endeavor Information Systems," http://www.exlibrisgroup.com/newdetails.htm?nid=504 (accessed January 5, 2007).

18. Ted Koppel, e-mail to author, January 15, 2007.

19. Collins, "Electronic Resource Management," 126.

20. Ibid., 127-139.

21. Ibid., 126.

22. Ibid., 127-139.

23. Ellen Finnie Duranceau, "Electronic Resource Management Systems, Part II: Offerings from Serial Vendors and Serial Data Vendors," *Against the Grain* 17, no. 3 (June 2005): 66.

24. Jewell "Electronic Resource Management Report," 142.

25. Stephen Meyer, "E-Matrix—Choosing to Grow Your Own Electronic Resource Management System," *Serials Review* 32, no. 2 (June 2006): 103.

26. Ellen Finnie Duranceau, "Electronic Resource Management Systems From ILS Vendors," *Against the Grain* 16, no. 4 (September 2004); Collins, "Electronic Resource Management"; Duranceau, "Electronic Resource: Part II"; Stephen Meyer, "Helping You Buy: Electronic Resource Management Systems," *Computers in Libraries* (November/December 2005); Geller, "ERM: Staffing."

27. Jewell, "Electronic Resource Management Report."

28. Dodie Ownes, "Findability Enabled," *Library Journal* 131, no. 13 (August 2006): http://www.libraryjournal.com/article/CA6359876.html (accessed December 15, 2006).

29. Rafal Kasprowski, "Recent Developments in Electronic Resource Management in Libraries," *Bulletin of the American Society for Information Science and Technology* 32, no. 6 (2006): http:www.asis.org/Bulletin/Aug-06/kasprowski.html (accessed November 5, 2006).

30. Geller, "ERM: Staffing," 12.

31. Collins, "Electronic Resource Management", 139.

32. John Gregory Mathews, e-mail to SERIALST discussion list, May 11, 2006.

33. Henry McCurley, e-mail to SERIALST discussion list, May 10, 2006; Norm Medeiros, e-mail to SERIALST discussion list, May 10, 2006.

34. Melissa Jean Holmberg, e-mail to SERIALST discussion list, May 11, 2006.

35. Medeiros, e-mail, May 10, 2006.

36. Medeiros, e-mail, May 10, 2006; Holmberg, e-mail, May 11, 2006; Matthews, e-mail, May 11, 2006.

37. Medeiros, e-mail, May 10, 2006.

38. Holmberg, e-mail, May 11, 2006; Matthews, e-mail, May 11, 2006.

39. Medeiros, e-mail, May 10, 2006.

40. Holmberg, e-mail, May 11, 2006.

41. Matthews, e-mail, May 11, 2006

42. Medeiros, e-mail, May 10, 2006; Matthews, e-mail, May 11, 2006.

43. Holmberg, e-mail, May 11, 2006; Matthews, e-mail, May 11, 2006.

44. Medeiros, e-mail, May 10, 2006.

45. Holmberg, e-mail, May 11, 2006; Matthews, e-mail, May 11, 2006.

46. Matthews, e-mail, May 11, 2006.

APPENDIX. Licensing Work Form for E-Matrix

License Name (Name of Physical File):

NCSU Contract Number: [ ] ; Filed in physical folder? [ ] Yes [ ] No

Licensor (Include address, e-mail, phone numbers):

Licensee (Include address, e-mail, phone numbers):

_____

_____

_____

_____

LOCKSS? [ ] Yes [ ] No   **Portico?** [ ] Yes [ ] No

License Execution Date: _____

License Start Date: _____   End Date: _____

License Replaces: _____

Contract Advisory Needed? [ ] Yes [ ] No

Reason for Contract Advisory: _____

Date sent: _____

Technical Contact (if different from Licensor):

_____

Cure Period for Breach ? _____ (Days, Weeks, Months)

Termination for Cause: Section No: _____

_____

Period for Termination: _____ (Days, Weeks, Months)

_____

Reason for Termination: _____

Intellectual Property Warranty? [ ] Yes, Section no: _____ ; [ ] No Content Warranty? [ ] Yes, Section no: _____ ; [ ] No

Limitation of Liability? [ ] Yes, Section no(s): _____ ; [ ] No Governing Jurisdiction: _____

Usage Rights

Fair Use Clause: ☐ Present, Section no.:_____   ☐ Absent

Nature of Use Defined? ☐ Yes, Section no.:_____   ☐ No

Print Copy:
☐ Permitted (explicit) ☐ Permitted (interpreted)
☐ Prohibited (explicit) ☐ Prohibited (interpreted) ☐ N/A
Note:_____

Scholarly Sharing:
☐ Permitted (explicit) ☐ Permitted (interpreted)
☐ Prohibited (explicit) ☐ Prohibited (interpreted) ☐ N/A
Note:_____

ILL Secure Electronic Transmission:
☐ Permitted (explicit) ☐ Permitted (interpreted)
☐ Prohibited (explicit) ☐ Prohibited (interpreted) ☐ N/A
Note:_____

Library Reserve, Print:
☐ Permitted (explicit) ☐ Permitted (interpreted)
☐ Prohibited (explicit) ☐ Prohibited (interpreted) ☐ N/A
Note:_____

Digitally Copy:
☐ Permitted (explicit) ☐ Permitted (interpreted)
☐ Prohibited (explicit) ☐ Prohibited (interpreted) ☐ N/A
Note:_____

ILL Print/Fax:
☐ Permitted (explicit) ☐ Permitted (interpreted)
☐ Prohibited (explicit) ☐ Prohibited (interpreted) ☐ N/A
Note:_____

ILL Electronic:
☐ Permitted (explicit) ☐ Permitted (interpreted)
☐ Prohibited (explicit) ☐ Prohibited (interpreted) ☐ N/A
Note:_____

Library Reserve, Electronic:
☐ Permitted (explicit) ☐ Permitted (interpreted)
☐ Prohibited (explicit) ☐ Prohibited (interpreted) ☐ N/A
Note:_____

Course Pack Print:
☐ Permitted (explicit) ☐ Permitted (interpreted)
☐ Prohibited (explicit) ☐ Prohibited (interpreted) ☐ N/A
Note:

Electronic Links:
☐ Permitted (explicit) ☐ Permitted (interpreted)
☐ Prohibited (explicit) ☐ Prohibited (interpreted) ☐ N/A
Note:

Concurrent Users (enter a number):
Note:

Remote Access? ☐ Yes ☐ No
ILL Record Keeping Requirement? ☐ Yes ☐ No
Archiving Right? ☐ Yes ☐ No ☐ Undetermined
Archiving Format: ☐ Remote ☐ CD-ROM ☐ Tape
Unspecified Tangible
Perpetual Access Right? ☐ Yes ☐ No
Note:

Course Pack Electronic:
☐ Permitted (explicit) ☐ Permitted (interpreted)
☐ Prohibited (explicit) ☐ Prohibited (interpreted) ☐ N/A
Note:

Permitted Access Note:

Other User Restrictions Note:

Confidentiality of User Information Indicator ☐ Yes ☐ No

Archiving Note:

☐ **Perpetual Access Holdings:**
Note:

Authorized Users

Authorized User Definition in Section no. _____

	Permitted (explicit)	Permitted (interpreted)	Silent
Registered Users (locally defined)	☐	☐	☐
Registered Borrowers	☐	☐	☐
Faculty	☐	☐	☐
Staff	☐	☐	☐
Students	☐	☐	☐
University Affiliates	☐	☐	☐
Distance Education	☐	☐	☐
Guests	☐	☐	☐
Administrators	☐	☐	☐
Unlimited	☐	☐	☐

Chapter 11

Integration and Data Standards

Mark Ellingsen

INTRODUCTION

In recent years, a suite of applications have been developed to accompany the traditional Integrated Library System (ILS) in enabling libraries to manage and provide access to information. These applications include metasearching and portal software; resolvers using the OpenURL syntax for context-sensitive linking; digital asset management; and electronic resource management (ERM) systems.[1] Lorcan Dempsey adds to this list, pointing to tools such as personal bibliographic and reading list software.[2] With this proliferation of systems of which the ILS is only a part, libraries face the issues of data standards and application integration. This chapter will explore these issues. The first part will look at the various standards in place for accessing and managing e-resources. The second part will focus on the integration of library applications and, in particular, the hopes invested in Web services technology.

AUTHENTICATION AND AUTHORIZATION

Many e-resource providers use authentication based on IP address ranges from subscriber organizations. This is a relatively common method to authenticate due to its ease of implementation. However, it also has some serious disadvantages, not least of which is its susceptibility to address spoofing in which incoming IP packets have forged addresses. Second, IP address ranges are associated with the network domain of a subscribing organization, which makes it impossible for bona fide users to access these resources from machines that are off campus. This method forces an institution to implement a Web proxy that connects to the e-resource site with an acceptable IP address on behalf of the end user—in effect, hiding the initial IP address from the vendor.

The obvious alternative to IP authentication is to provide authentication via a username and password. The problem with this authentication method is that the user must remember a username and password combination for each e-resource provider. One way around this problem is for providers to sign up to a single username and password authority. This was the thinking behind the Athens authentication service, which has had widespread take-up in the United Kingdom and a number of other countries. In the past few years, the Athens service has allowed for local authentication by calling an Application Programming Interface (API) or by using Security Assertion Markup Language (SAML), which is an XML standard for exchanging information about identity, authentication, and authorization information (http://www.oasis-open.org/committees/tc_home.php?wg_abbrev=security). It is also the key standard underpinning the Shibboleth federated identity infrastructure (http://shibboleth.internet2.edu/shib-faq.html), which is being adopted in a number of countries such as the United States, United Kingdom, Australia, Switzerland, and the Netherlands. Shibboleth is based on trust between identity providers and service providers who come together in a federation. A potential user of an e-resource must be registered and authenticated with an identity provider. Once the user attempts to access an e-resource, a number of messages are exchanged in which the service provider requests both proof of authentication as well as attributes of the user from the identity provider. This allows the service provider to make decisions about authorization. The attributes that make up the identity, class, and role of the user are commonly held within a directory service (based on the Lightweight Directory Access Protocol [LDAP]) or relational database.

The last group of authentication and authorization standards that needs to be looked at are those pertaining to accessing a library's e-resources within the institution. Today one would expect any new application to integrate with the authentication mechanisms used by the institution where that is possible. Authentication is usually handled through a directory service built upon LDAP and using a protocol such as Kerberos. Furthermore, it is now expected that Web applications should integrate with single sign-on mechanisms such as CAS (http://www.ja-sig.org/products/cas/) and Pubcookie (http://www.pubcookie.org/). At the very least, it should be possible to integrate an ERM system with a directory service, and one would hope that vendors will provide integration with the popular Web single sign-on (SSO) packages if the ERM system is a Web application. The integration with Shibboleth would also be of benefit to consortia where the management of e-resources is done both at a local level and a consortial level. In this scenario, members of the consortium would belong to a Shibboleth federation in which authentication was done at the local institution and attributes would be sent to the ERM system for arriving at authorization decisions.

SEARCH AND RETRIEVE

Authentication and authorization are, of course, only one aspect of accessing a resource. Finding and retrieving relevant items is the next stage in that process. Traditionally, this has been done by using the interface native to a resource to search for relevant items. The alternative has been to use integrated library systems software or bibliographic reference software as a tool to search other bibliographic databases using the Z39.50 protocol. These methods predate the World Wide Web, with the Z39.50 protocol having been initially developed in the 1980s. With the development of the Internet, the rise of the Web and the increasing amount of resources being made electronically, the variety and number of interfaces has risen dramatically.

The response to this profusion of interfaces has been the development of metasearch engines. Typically, a connection from a metasearch engine to a bibliographic database has been via the Z39.50 protocol. In many instances, however, the exchange of information has instead been via the target service's API. Partly, this is due to the perceived functional complexity of the protocol, but in the main this is because the protocol predates the Web and, as such, does not use HTTP. With the proliferation of Web interfaces, vendors have quite understandably provided Web APIs. Many have not invested in providing a Z39.50

interface. Even when the protocol has been implemented this has not been done in a consistent manner across all Z39.50 enabled services. For the most part, this is because the protocol abstracts from the actual implementation of a search. Therefore, for example, the implementation of an author search is dependent upon the specifics of the data model and associated indexes. This may vary from service to service leading to inconsistent result sets from similar database services. In an attempt to deal with this situation, the U.K. Joint Information Systems Committee funded a project in 1999 to provide an application profile, known as the Bath profile (http://www.collections canada .ca/bath/ap-bath-e.htm), to bring some standardization into implementations of Z39.50 for library systems. However, as Nicolaides[3] discusses, take-up by bibliographic database vendors has been slow. Recently, however, the Bath profile has been used as a foundation by the U.S. national profile NISO Z39.89. Given the standing of the National Information Standards Organization (NISO) and the dominance of U.S. vendors in the library software market, it is likely that this will supersede the Bath profile at some point in the future.

Regardless of the new application profiles, Z39.50 is still a pre-Web protocol. The response of the Z39.50 community to the growth of the Web was the Z39.50: Next Generation (ZNG) initiative. A group of Z39.50 implementers met in 2001 to develop a search and retrieval protocol that utilized HTTP. The outcome of this initiative was two new protocols: Search/Retrieve via URL (SRU) and the Search/Retrieve Web service (SRW). These utilize a new query language known as the Common Query Language (CQL). CQL queries can be set within a context or, in the jargon of CQL, a "context set." For example, there is a context set for the Z39.50 Bath profile in which access points or search categories such as the "Uniform Title" can be used within the framework of CQL. This work has been built upon the functionality of Z39.50 but with a lot of the complexity removed in the hope that this will encourage a wider take-up. The difference between the two protocols is that SRU is purely HTTP based while SRW uses a messaging framework called SOAP for transporting and routing XML documents. This work is now under the auspices of the Library of Congress.

One other piece of work to come out of the ZNG initiative was Z39.50 Explain Explained and Re-Engineered in XML (ZeeRex). As the name might suggest, this was an attempt to repurpose for the Web the Z39.50 explain mechanism, which is used for describing Z39.50 databases, and to make it simpler. This has now been taken up by the NISO Metasearch Initiative as a basis for the draft Z39.92-200x, *Information Retrieval Service Description Specification.*[4] Originating from a 2003 American Library Association (ALA) meeting, the Metasearch Initiative was a response to the lack of standardization in searching and retrieving results for metasearch or federated search engines. The problem of varying implementations of Z39.50 was compounded by the variety of other technologies, such as XQuery and proprietary APIs, now being used. There are three task groups within the Metasearch Initiative: one focuses on access management, another focuses on collection and service descriptions, and the third group focuses on search and retrieval. In addition, to the report noted in previous text, the collection and service description task group has also produced a draft specification for the description of collections.[5] The access management group produced a report in 2005 on "Ranking of Authentication and Access Methods Available to the Metasearch Environment" in which ease of use and environmental factors of different authentication methods were evaluated.[6] The search/retrieve working group has produced an implementer's guide to the NISO Metasearch XML Gateway (MXG) protocol.[7] The MXG protocol helps content providers to return a standardized XML response to a query from a metasearch engine. This protocol, which is discussed in detail by Hodgson, Pace, and Walker,[8] is based on the SRU protocol and has three levels of implementation of which only level three is fully SRU compliant.

Given the earlier discussion of ongoing initiatives, it is obvious that standards to enable access to e-resources are very much still works in progress and that even the relatively stable standards such as Z39.50 continue to evolve. As we have seen, the rise of new standards and changes to existing ones occur because of shifting technological paradigms, or it becomes obvious that existing standards are not as helpful as originally intended. This makes it difficult both for library system vendors as well for those who are writing a request for proposal (RFP) as part of a procurement exercise. One can only ask vendors to develop systems based on mature standards and to incorporate the newer ones when the time is appropriate. The latter is very much dependent on the extensibility of the underlying software architecture.

Before concluding this section, it is appropriate to mention two other standards that have now reached a certain maturity and have become ubiquitous in implementation for products in the e-resource marketplace. These are OpenURL (http://www.niso.org/standards/standard_detail.cfm?std_id=783) and OAI-PMH (http://www.open archives.org/OAI/openarchivesprotocol.html). OpenURL (Z39.88) is a standard for encoding e-resource metadata within a URL. Link resolvers use this metadata to provide context sensitive services pertaining to that resource including links to "appropriate copies" of full-text based on user affiliation and permissions. Link resolvers provide services at the conclusion of a bibliographic search and use a knowledge base to determine the institution's access to e-resources. For this reason, it is a service complimentary to a search engine. For additional discussion on the OpenURL standard, see Apps and MacIntyre.[9]

The second standard, the Open Archives Initiative Protocol for Metadata Harvesting (OAI-PMH), is a protocol for harvesting metadata from digital repositories. These metadata records, which are usually encoded using Dublin Core, can then be searched for relevant material with link servers providing the link back to the digital document. As the growth of digital repositories increases, this protocol will become even more important.

APPLICATION INTEGRATION

The issues surrounding application integration are neither new nor unique to libraries. Indeed, the IT industry has spent many years building integration solutions, worrying over the complexities in doing so, and attempting to provide standards which can make integration easier. IT in libraries is no exception. Library system vendors have been dealing with integration standards for many years. They have successfully provided interfaces to book and serial vendors using electronic data interchange (EDI) and have interfaced self-issue terminals to the library system using Session Initiation Protocol (SIP) and NISO Circulation Interchange Protocol (NCIP). Some are now beginning to incorporate the newer Web service standards into their products such as the ERM system. For example, the ERM system from Ex Libris, Verde, provides a Web service SOAP interface. However, before discussing Web services, it would be useful to look at other integration techniques for comparison.

Most vendors provide a utility to upload a text file of student records into the ILS. Known as Extract, Transform and Load (ETL), the process extracts data from the student record system, transforms it into the correct format for the upload utility including any change of values, and then loads it into the ILS data store. This is a common technique in populating data stores where a duplication of data is required and where a change to the source data between each load is not critical. Systems librarians often have to write the extraction and transformation scripts but this is not atypical compared to other integration techniques. This mode of integration suits the delivery of usage statistics and licensing information for ERM systems because changes are relatively slow. However, conversely, it is not suitable for real-time updates or where updates are part of a larger process with embedded business rules.

A second method of integration in widespread use is tying things together through unmediated access to data stores. From the 1990s onward, the use of Relational Database Management Systems (RDBMS) became pervasive and libraries were keen to exploit this technology to provide better data integration between the library system and other applications. It certainly made the extraction of data from the library system an easier task because there is a well-published standard for the extraction, insertion, and manipulation of data within a relational database, the SQL language. However, there was an expectation amongst many that, once the data was open and no longer hidden behind a proprietary application, it would be a lot easier to provide data integration in real time. Unfortunately, this has proven difficult to achieve. This is not due to the technology; one can quite easily add database triggers that fire when a data change occurs in one database and then write that change to another database. The problem lies with the fact that accessing the data model directly usually circumvents the business logic, rules, and constraints embedded within the application layer. There is a real danger that directly updating a data element may introduce an inconsistency into the database if related data elements are not also appropriately updated.

In order to ensure that changes to the data are consistent, the developer must write to an API. An API hides the complexity of the internal workings of the application, business logic, and data model from the developer. As long as the functions provided by the API remain consistent across versions of the application, the developer has no need to be concerned that the application, business logic, or data model has changed. The developer only needs to show an interest in any new features that the API may provide. So, if the developer uses the API, any updates to the underlying data will be handled in a manner consistent with the intentions of the vendor's application developers. However, the developer has to learn the specific technology that the API utilizes, but, as we shall see, Web services technology provides a standard for this.

Furthermore, even if one understands the business logic and therefore can be confident that making direct changes to the data will not introduce inconsistencies, one has to be aware of any changes to the data model through upgrades and deal with these accordingly. Changes to the model often occur when new functionality or extra pieces of data are added by the system vendor. This can lead to a growing maintenance problem as changes to the integration code need to be tested at every upgrade. It should also be borne in mind that suppliers are often reluctant to provide support if any inconsistencies are introduced by circumventing their application.

WEB SERVICES

If the lack of standards and/or the inherent complexity of application integration have stood in the way of any real progress toward the integration of library systems with other corporate systems, then many have argued that Web services hold the promise that this sort of integration may at last be on the horizon. A narrow definition of a Web service is a software service that can process and act upon a message contained in an XML document that it receives via transport and application protocols. There is a lack of consensus within the industry as to any broader definition than this and, indeed, as we shall see, there are at least two opposing camps. Nevertheless, information professionals have been talking about the potential of Web services in the library domain for a number of years now. In 2002, Cordeiro and Carvalho[10] argued that "Web services technology can ease substantially the integration of applications from different vendors, in a 'plug and play' mode . . . such a technology appears to be a very promising way to overcome the constraints of the traditional 'monolithic' library automation systems." Web services reached the Library and Information Technology Association's top ten technological trends to watch out for in both 2003 and 2004.[11]

Commentators see a number of benefits to adopting Web service technology. In particular, Web services are platform independent and vendor neutral; they are based on standards to which all vendors can sign up to; they are loosely coupled, so that there are no unwanted dependencies; and they can be anywhere on the Internet and can be addressed with a URI, of which URLs are a subset. For these reasons, together with major vendor buy-in, there is much excitement and hype about Web services. However, it should also be noted that Web services have some disadvantages. For example, they are slower compared with other distributed computing environments due to the need to parse an XML text file.

There are two competing philosophies for Web services. An awareness of the basic differences between these philosophies is important, if only because library system vendors will mention at least one of these as part of their suite of integration standards. The dominant philosophy, at least amongst IT vendors, is based around SOAP (formerly this acronym stood for Simple Object Access Protocol, but SOAP's evolving role has rendered this name misleading and it has now been dropped). As mentioned earlier, SOAP is a messaging framework that provides a standard message format in which an XML document is contained within a message body. There may also be an optional header, which may include metadata about the message and routing information. Both header and body are contained within what is known as the SOAP envelope. Note also that SOAP-based Web services are not dependent on HTTP and can use other application protocols such as the Simple Mail Transfer Protocol (SMTP)—in other words, they can be sent by e-mail. The Web service itself is described using XML in the form of the Web service Description Language (WSDL), which describes the public interface to the service. The Web service may also be registered in a directory, for discovery purposes, using the Universal Description, Discovery and Integration (UDDI) specification, to provide access to the WSDL documents. For a more detailed discussion of SOAP-based Web services and their implications for library systems see Cordeiro and Carvalho.[12]

The competing philosophy of Web services is known as Representational State Transfer (REST). Originally elucidated by Roy Fielding,[13] REST is a software architectural style. Proponents of this philosophy of Web services, such as Prescod,[14] argue it underlies the architectural framework of the World Wide Web. REST describes the process of transferring state between systems over the Web, in a similar fashion to that of a Web browser and Web server. Each piece of information, known as a resource, is uniquely addressable by a URI, in much the same way as a Web page. Representations of the resource, which are usually XML files, are exchanged between the client and the server. These representations are accessed and manipulated using the HTTP methods GET, POST, PUT, and DELETE. The resources themselves are never accessed directly through HTTP, and only the application fronted by the Web service has direct access to the resource. The client moves from state to state and from one resource to another based on a transition from one URI to another. Its proponents argue that there is no need for the additional abstraction of SOAP.

Commentators such as Snell,[15] Hinchcliffe,[16] and Prescod[17] present detailed arguments as to the relative merits of the two philosophies. Although this is not the place to discuss these arguments, it is worth noting some of the implications for application integration and the impact these may have on developers employed by the institution to implement integration projects. SOAP-based Web services are more complex than REST-based Web services and therefore require developers to possess additional skills in this field. Any integration between applications that takes place in-house needs to consider the added complexity of SOAP. However, because it may be more difficult to redesign a legacy system to use a REST-style interface, SOAP may be unavoidable for wrapping legacy systems with a Web services layer. The implication is that effort needs to be expended into ensuring that consumers of the SOAP service must align with the SOAP implementation at the server end. This means that REST has the edge with regard to interoperability between heterogeneous platforms. However, it may well be that the service needs to implement features which are not well supported within REST, such as asynchronous event notification or the orchestration of interactions between multiple services.

The choice of which to use, SOAP or REST, is dependent upon a number of factors, and ultimately the vendor or service provider is the only actor knowledgeable enough about their systems to make that call. It may be of interest that service providers such as Amazon provide both SOAP and REST-based interfaces with the latter being far more popular. The FEDORA digital object repository also provides both SOAP and REST interfaces.

Of course, whatever the choice, integration does not come free. One still needs experienced developers to make that integration a reality. However, the point here is that when procuring an ERM system or other library software, it is important to ensure that the application's API is built around Web services or has a Web service wrapper. Without this, developers need to learn the technology that the API is based on rather than just work with an XML file or SOAP message. In addition, it would be more difficult to integrate the application into any Enterprise Application Integration (EAI) solution based around Web service standards. EAI is about bringing together heterogeneous distributed systems to support business processes and data sharing. Where business processes are dependent upon multiple systems, it may be necessary to build an application that supports a workflow and data flow between systems. This is where EAI solutions come into play, and currently they are closely based around Web service standards such as the Business Process Execution Language (BPEL).

SERVICE-ORIENTED ARCHITECTURE (SOA)

Up to now, the discussion has focused on the technologies and standards required for integrating applications. We have seen that there are good reasons for adopting Web services. However, this is not enough. Without an understanding of the business processes that might go on within an organization, the risk is that any API, whether or not it is based on Web services standards, will be of limited use. By themselves, Web services do not solve the problem of integration; instead, they are an enabling technology. It is only by analyzing business processes and developing APIs to support these processes that integration becomes possible. The library world has had plenty of experience providing integration solutions, and business processes are well understood, so this is nothing new. One example is the use of the protocols

SIP and NCIP to exchange circulation messages or the use of EDI to order items from book vendors. These services are narrow in definition; hence, the application protocols that are built upon them are well-defined.

However, the library IT industry faces at least two major integration challenges. The first is the fragmentation of library management software and the need to integrate applications between multiple vendors; the integration of ERM systems with library management systems is a case in point. For example, some form of integration between an ERM system and the acquisitions and serials modules of an ILS is essential if serials librarians are not to make the mistake of cancelling a print subscription that has given the library advantageous terms for the electronic subscription. This type of integration should be feasible if the spirit of cooperation wins over the instinct for competition between library software vendors. There are already examples of such cooperation—for instance, the partnership between Endeavour and Talis in integrating Endeavour's Meridian ERM system with Alto, the ILS offering from Talis.

The second challenge is the continuing need to integrate with other applications in use within the institution. For example, not only should the ERM system be integrated with the acquisitions module of the ILS but both should have some form of integration with the institution's finance system. However, this is dependent upon the development and use of standards such as the Universal Business Language (UBL). So, for example, one can imagine a process by which an order for an e-resource package may be initiated within the ERM system, which then updates the ILS acquisitions and serials modules and at the same time communicates with the finance system to create an order. Of course, the technical implementation of this may not be as straightforward, as there may have to be brokers between systems to do message translation, handle data flow, and embed some business logic. However, if the messages are based on UML documents, then there is a possibility that this might be feasible.

The challenge of integrating library systems with other corporate systems brings us to the latest trend in application architecture, which is known as service-oriented architecture (SOA). The thinking around SOA is that we should be building services that can be loosely coupled and reused in a variety of contexts. These services model coarse-grained business functionality. This will have profound implications for software design and, if adopted, will lead to the breakup of monolithic systems into smaller services. In theory, this should allow an ERM system or an ILS to take advantage of an "order creation" service rather than replicate ordering functionality within the application itself. A lot of work on SOA is occurring under the auspices of the Organization for the Advancement of Structured Information Standards (OASIS). For example, OASIS is working on "developing, publishing and maintaining archetypal 'blueprint' sets of requirements and functions to serve as generic, vendor-neutral instances of service-oriented solutions for real business requirements."[18] This move toward service orientation has been recognized within the library community. The Digital Library Federation has established a project to look at a "service framework."[19] The U.K. Joint Information Systems Committee (JISC) and Australia's Department of Education, Science and Training (DEST) have cooperated on the e-Framework for Education and Research.

The e-Framework supports a service oriented approach to developing and delivering education, research and management information systems. Such an approach maximizes the flexibility and cost effectiveness with which systems can be deployed, both in an institutional context, nationally and internationally. [Furthermore,] By documenting requirements, processes, services, protocol bindings and standards in the form of "reference models" members of the community are better able to collaborate on the development of service components that meet their needs (both within the community and with commercial and other international partners).[20]

This initiative recognizes that, if we are to move forward, then these initiatives must also have vendor participation. In addition, these initiatives must extend beyond the library domain if library software is to integrate with other corporate systems.

SERIALS INFORMATION AND USAGE STATISTICS

Much of what we have discussed so far has been concerned with the development of integration technologies and architectures and the response of library software vendors and the library community engaged in IT concerns. Some of these technologies, such as SOAP-based Web services, are now being incorporated within newer products such as ERM systems. Some of the recent work in messaging between serial vendors and library software vendors can take advantage of these Web service technologies.

The ONIX for Serials messaging formats (http://www.editeur.org/onixserials.html) are of direct relevance for ERM systems. These facilitate the exchange of serials information within the serials industry and with libraries. There are three message formats: serial online holdings (SOH); serial products and subscriptions (SPS); and serial release notification (SRN). These were developed in the past few years under the auspices of EDItEUR and NISO with a host of collaborators including serials publishers, hosting services, library system vendors, and university and national libraries. The messages are expressed in XML with appropriate validation schemas and DTDs (document type definitions). The SOH format allows the exchange of data with regard to the holdings of online hosting services. It can, for example, be used to populate a local knowledge base for use by the ERM system. This can be the same knowledge base used by other software such as link resolvers. The SPS format facilitates the exchange of availability and pricing information including how access to an electronic copy is affected by purchase of the printed format. The SRN format will allow for the exchange of article and issue level information and should facilitate electronic check-in.

The Digital Library Federation has also been very proactive in encouraging the development of standards in this area. The first phase of its Electronic Resource Management Initiative ended in August 2004 with a report that has been very influential.[21] The initiative produced an XML schema for encapsulating license data as a proof of concept. This work was conducted with awareness that there would be some synergy with work going on in the area of rights expression languages such as the Open Digital Rights Language (http://odrl.net/) and the Creative Commons RDF open access and fair use schema (http://creativecommons.org/). However, neither of these groups provided a sufficient basis for an ERM license expression schema. Therefore, the initiative decided to concentrate on providing its own schema based on a developed data dictionary of more than 300 elements and a data structure. The second phase of the initiative was inaugurated in

2006 with the following goals: to provide a rigorous review of the data dictionary to make it extensible and more coherent; to relate the work on license expression to the ONIX for Licensing Terms project; and to provide more focus on usage statistics, in particular to work with the SUSHI project (http://www.diglib.org/standards/dlf-erm05 .htm).

The ONIX for Licensing Terms (OLT) project was begun in 2005 with support from EDItEUR and from the U.K. Joint Information Systems Committee (http://www.jisc.ac.uk/) and the Publishers Licensing Society (http://www.pls.org.uk/ngen_public/default.asp). In addition, EDItEUR, NISO, DLF, and PLS established the License Expression Working Group to "to develop a single standard for the exchange of license information between publishers and libraries."[22] This will ensure that the work being carried out by the DLF's ERM initiative will be incorporated into the OLT project. One of the explicit objectives is to allow a publisher's licensing terms to be loaded into an ERM system. The first message ONIX-PL (Publisher License) will be: "an XML message format that can deliver a structured expression of a publisher's license for the use of (digital) resources, from publisher to agent to subscribing institution (or consortium)."[23] This will include "a specification, an XML schema, and a formal dictionary of controlled values."[24] There has also been some discussion on developing tools to allow publishers and libraries to produce OLT license expressions.[25] It is important to note that this is a work in progress and it will be a while before we see the routine communication of licensing terms between systems delivered in the new standards. Current information on the OLT project can be found at the OLT site (http://www.editeur.org/onix_licensing.html).

The other area in which standards have been progressing is in usage statistics. Project COUNTER (Counting Online Usage of Networked Electronic Resources) was launched in 2002 to investigate ways of standardizing online usage statistics for electronic resources (http://www.projectcounter.org/). The project has since released two codes of practice that provide standards and protocols for recording usage data. There are separate codes of practice for books and reference works and for journals and databases. A number of vendors now comply with these codes of practice, some even providing the statistics in the form of XML documents. For example, with respect to e-journals, there are two reports that vendors must supply. The first is the number of successful full-text article requests by month and journal. The second is the number of "turnaways" (i.e., rejected sessions) by month and journal. There are also a couple of optional reports that document the number of successful item requests and turnaways by month, journal, and page type and second, total searches run by month and service. However, while COUNTER defines standard usage data it does not help with communicating that data between machines. Currently, data often comes in the form of Excel spreadsheets. Project SUSHI (Standardized Usage Statistics Harvesting Initiative) aims to provide a standardized container to allow for the automatic transfer of COUNTER data from a vendor into an ERM system or other repository. Like many other projects, this also comes under the auspices of NISO. Initially, this will be a request/response model based on a Web services layer. As of October 2006, a number of vendors are participating, including EBSCO Information Services, Ex Libris, Innovative Interfaces Inc., and SWETS Information Services. For additional information about the SUSHI project, see the presentation given by Tim Jewell and Oliver Pesch at the midwinter conference of the American Library Association.[26]

USER INTERFACE INTEGRATION

The earlier discussion has concentrated on technologies for application integration at the level of data flow and business processes. However, there may be occasions when what is required is integration between user interfaces. There are at least two important technologies that should be considered with regard to user interfaces. First, Web feeds based on the RSS and Atom technologies may be useful to alert certain portions of a library's user community of new packages, new e-resources for trial, and so forth. Web feeds are already being used quite extensively within universities, and it may be useful for ERM systems to create news feeds that build upon this widespread use.

Second, we must consider the technologies involved in the integration of applications within a portal framework. Portals bring together multiple applications within a single user interface, each application being rendered within an area of the portal known as a channel or portlet. More formally: "a portlet is a user-facing, interactive application component that renders markup fragments that can be aggregated and displayed by a portal."[27] Portal software such as uPortal, Apache Jetspeed, and Oracle Portal are being adopted by institutions as a way to ease user navigation between different applications. The markup fragments, in HTML or XHTML, are usually the product of a transformation of the data stream from the source application. The original data stream is often an XML document, which is then transformed into the Web page markup via the XLST language. Having the original output as XML makes it easier for the developer to render the final markup, which is one more reason why vendors should provide a Web service API to their products. However, it is not just a matter of rendering some information into the portlet. Ideally, the portal users should be able to interact with the remote application that is feeding the portlet. The Web Services for Remote Portlets (WSRP) standard (http://www.oasis-open.org/committees/download.php/11774/wsrp-faq-draft-0.30.html) allows the developer to deploy a remote portlet in a portal with the use of Web services technology to handle the communication between them.

CONCLUSION

In this chapter, we have looked at the changing landscape of IT standards related to e-resources and application integration. These standards are ever-evolving and occasionally they are overthrown by a new computing paradigm. The challenge of new ways to utilize the power and ubiquity of the Web has led to an overhaul of the previous standards and an introduction of new ones. This is a fast changing arena and one in which libraries and software vendors need to be agile in order to take advantage of the new opportunities in e-resource management. Second, this chapter has examined some of the issues relating to application integration. The increasing multiplicity of library applications and the increasing importance of having software that can support business processes which may cut across traditional domains requires that library applications have rich and well supported APIs. Web services have become an important technology in the integration of services. Library software vendors need to ensure that these technologies are supported by their product suite. Unfortunately, integration does not come cheap. Integration between applications requires effort and investment from the institutions themselves both in the employment of developers and in the procurement of infrastructure such as EAI technology.

NOTES

1. A. Felstead, "The Library Systems Market: A Digest of Current Literature," *Program: Electronic Library and Information Systems 38, no. 2 (2004):* 88-96, http://www.emeraldinsight.com/Insight/ViewContentServlet?Filename=Published/EmeraldFullTextArticle/Articles/2800380201.html.

2. L. Dempsey, "The Integrated Library System That Isn't" (February 22, 2005), http://orWeblog.oclc.org/archives/000585.html.

3. F. Nicolaides, "The Bath Profile Four Years On: What's Being Done in the UK?" *Ariadne* 36 (2003), http://www.ariadne.ac.uk/issue36/bath-profile-rpt/.

4. National Information Standards Organization, *Information Retrieval Service Description Specification* (NISO Press: Bethesda, Maryland, 2005), http://www.niso.org/standards/resources/Z39-92-DSFTU.pdf.

5. National Information Standards Organization, *Collection Description Specification* (NISO Press: Bethesda, Maryland, 2005), http://www.niso.org/standards/resources/Z39-91-DSFTU.pdf.

6. NISO Metasearch Initiative Task Group 1, *Ranking of Authentication and Access Methods Available to the Metasearch Environment* (NISO: Bethesda, Maryland, 2005), http://www.niso.org/standards/resources/MI-Access_Management.pdf .

7. NISO Metasearch Initiative Task Group 3, *Metasearch XML Gateway Implementers Guide* (NISO: Bethesda, Maryland, 2006), http://www.niso.org/standards/resources/MI-MXG_v1_0.pdf.

8. C. Hodgson, A. Pace, and J. Walker, "NISO Metasearch Initiative Targets Next Generation of Standards and Best Practices," *Against the Grain* 18, no. 1, (February 2006), http://www.niso.org/pdfs/ReprintNISOv18-1.pdf.

9. A. Apps and R. MacIntyre, "Why OpenURL?" *DLib Magazine* 12, no. 5 (May 2006), http://www.dlib.org/dlib/may06/apps/05apps.html.

10. M.I. Cordeiro and J. de Carvalho, "Web Services: What They Are and Their Importance For Libraries," *Vine* 32, no. 4 (2002): 46-62, http://www.emeraldinsight.com/Insight/viewContentItem.do?contentType=Article&contentId=862498.

11. Library and Information Technology Association (2003), http://www.lita.org/ala/lita/litaresources/toptechtrends/midwinter2003.htm. Library and Information Technology Association (2004), http://www.lita.org/ala/lita/litaresources/toptechtrends/annual2004.htm.

12. Cordeiro and Carvalho, "Web Services," 46-62.

13. Roy Fielding, *Architectural Styles and the Design of Network-Based Software Architectures,* University of California, Irvine Doctoral Dissertation (2000), http://www.ics.uci.edu/~fielding/pubs/dissertation/top.htm.

14. P. Prescod, "Second Generation Web Services," (2002), http://Webservices.xml.com/pub/a/ws/2002/02/06/rest.html.

15. J. M. Snell, "A Quick Look at the Relationship of REST-Style and SOAP-Style Web Services," (October 12, 2004), http://www-106.ibm.com/developerworks/webservices/library/ws-restvsoap/.

16. D. Hinchcliffe, "REST vs. SOAP: Battle of the Web Service Titans—Picking a Winning Web Service Approach," SOA Web Services Journal (April 26, 2005), http://Webservices.sys-con.com/read/79282.htm.

17. P. Prescod, "Second Generation."

18. Organization for the Advancement of Structured Information Standards, "OASIS SOA Adoption Blueprints TC," http://www.oasis-open.org/committees/tc_home.php?wg_abbrev=soa-blueprints.

19. Lorcan Dempsey and Brian Lavioe, "DLF Service Framework for Digital Libraries: A Progress Report for the DLF Steering Committee" (2005), http://www.diglib.org/architectures/serviceframe/dlfserviceframe1.htm.

20. Joint Information Systems Committee and Australia's Department of Education, Science and Training, *The E-Framework for Education and Research: A QA Focus Documents* (2006), http://www.ukoln.ac.uk/qa-focus/documents/briefings/briefing-91/briefing-91-A5.doc.

21. Timothy Jewell et al., *Electronic Resource Management: Report of the DLF ERMI Initiative* (Washington, DC: Digital Library Federation, 2004), http://www.diglib.org/pubs/dlf102/ (accessed December 2, 2006).

22. National Information Standards Organization, "License Expression Working Group," http://www.niso.org/committees/License_Expression/LicenseEx_comm.html.

23. Presentation by David Martin to UKOLN in April 2006, http://www.ukoln.ac.uk/events/drm-2006/presentations/d-martin.ppt.

24. Ibid.

25. See http://www.bic.org.uk/ppt/060705francis_OLT.ppt.

26. Tim Jewell and Oliver Pesch, "SUSHI: Standardized Usage Statistics Harvesting Initiative," Presentation at ALA Midwinter January 2006, http://www.niso.org/committees/SUSHI/SUSHIpresentationALA2006.pdf.

27. OASIS, Web Services for Remote Portlets 1.0 Frequently Asked Questions, http://www.oasis-open.org/committees/download.php/11774/wsrp-faq-draft-0.30.html.

Chapter 12

E-Journal Management Tools

Jeff Weddle
Jill E. Grogg

INTRODUCTION

E-journal management tools have become increasingly sophisticated and user-friendly, both for librarians and patrons. Initially, early tools from subscription agents and serials vendors, such as EBSCO Online (now EBSCO*host* Electronic Journals Service), were virtually all that existed in the commercial arena to assist librarians in the management of exponentially growing e-journal collections and assist users in finding the e-materials they desired. These serial vendor tools, however, were quickly joined by other services, such as A-to-Z lists, link resolvers, federated searching products, and MARC record services.

With the development of these and other e-journal management resources, subscription agents and serial vendors have been forced to reexamine their initial products and reevaluate their roles in e-journal management. The result for librarians is unprecedented access to a suite of symbiotic products; for example, the choice of an A-to-Z journal list inevitably affects the decision to purchase a link resolver product, which in turn affects the choice of a MARC record service. To further complicate matters, librarians must assess the interoperability of these disparate tools with their current or future Integrated Library System (ILS) and electronic resource management system (ERMS).

In short, over the past decade, librarians have seen a veritable explosion of electronic resource management (ERM) tools. Much like the modules in an ILS are interrelated, most of the e-journal management tools are interconnected through a common knowledgebase or a single point of data entry. Furthermore, for many libraries, e-journal management tools are no longer optional, any more than an ILS is optional. Therefore, librarians are faced with an increasing array of choices and associated tasks such as the following: whether to purchase all tools from the same vendor, whether to build a homegrown solution, whether to focus on interoperability with a current ILS and/or ERMS, how best to train staff and subsequently encourage staff to accept new roles, and finally, how to artfully adapt traditional technical services processing and workflow.

Issues of workflow are paramount, as most of the tools discussed in the following text are administered by varying departments and positions, depending on the library. Libraries have absorbed the administration of e-resource management tools in different ways, and, as these tools become more commonplace and central to meeting the mission of the library, librarians must reevaluate how best to distribute e-journal management duties. Ultimately, the near-ubiquity of e-journal management tools reflects a basic truth: the primary mode of access for many patrons has shifted from print to electronic and the library must accommodate this shift or risk marginalization.

While it is beyond the scope of this chapter to tackle all issues associated with e-journal management tools, the authors will specifically examine the development and deployment of A-to-Z journal lists, link resolvers, and federated searching products. MARC record services, subscription agent and/or serial vendor tools, and ERMS will be discussed only briefly; other chapters in this book, particularly Chapter 10, "ERM Systems," address these e-journal management tools in more detail.

A-TO-Z LISTS

E-journal lists, or A-to-Z lists, represent one of the first ways in which libraries attempted to create access points for e-journals. Initially, many librarians maintained their own Web lists for their e-journal collections—collections that were, at the time, relatively small. As the number of e-journals grew and the number of sources from which libraries could acquire access to e-journals increased, these sorts of "high maintenance Web lists"[1] became too costly and complicated for individual libraries to maintain. For example, the full text of *Journal of Black Studies* is currently available from the following access platforms and within the following timeframes:

- 09/01/1970 to 11/30/2002: JSTOR
- 01/01/1997 to present: Westlaw for Law Schools, Westlaw for Non-Law Schools
- 01/01/1997 to 11/01/1998: A number of Gale products, such as Academic OneFile
- 1998 to present: SwetsWise Online Content
- 01/01/1999 to present: EBSCO*host* EJS, Highwire Press, Sage Publications
- 03/01/2001 to 07/01/2002; 11/01/2002 to 07/01/2004; 11/01/2004 to present: Black Studies Center

It is important to note that as impressive as the preceding list is, it is not complete. For example, the entry for Gale has been collapsed into a general statement ("a number of Gale products"); in actuality, the *Journal of Black Studies* is available in more than five discrete Gale products. Regularly tracking and updating the specific dates and existence of journals both at the vendor (i.e., Gale) and resource (i.e., Academic OneFile) level consumes enormous amounts of staff time. Hence, as many librarians began to understand that maintaining their own A-to-Z journal lists was not scalable, companies such as Serials Solutions emerged. This company promises to "deliver tools and services that empower librarians and enable their patrons to get the most value out of their electronic serials, including content in aggregated databases, publisher Web sites and subscription agents" (http://www.serialssolutions.com/home.asp).[2]

Serials Solutions ostensibly tracks the known universe of all providers, including dates, for a predetermined number of electronic publications. The data collected populates a global knowledgebase, which is the cornerstone of the A-to-Z journal list. When a library purchases services from Serials Solutions, this library makes a copy of the global knowledgebase in its own image. In other words, a librarian literally checks off the e-journals to which the library has access—be they direct from publishers, through third parties such as Ingenta, or via aggregators—and creates a localized version of the global knowledge-base. Thus, the ability to provide a searchable A-to-Z journal list is born, and the burden of tracking which journals may fall in and out of which packages is removed from the librarian and placed on the service provider. While the local librarian must maintain his or her local knowledgebase, he or she is not required to know when and if a publisher removes its content from a particular aggregator or third-party.

While Serials Solutions was among the first companies to provide this sort of service, it is by no means alone in the current marketplace. A-to-Z journal listing services are now available from a number of providers, such as EBSCO (a traditional subscription agent); Ex Libris (an ILS vendor and provider of other tools such as SFX, a link resolver; MetaLib, a federated search tool; and Verde, an ERMS); and TDNet (an e-resource management company).

From their humble beginnings as literal lists, A-to-Z listing services have become more robust. E-journals can now be categorized and searched by subject. In addition, these listing services can include print holdings. By including print holdings in the e-journal lists, users are no longer required to conduct duplicate searches in various places, such as the Online Public Access Catalogs (OPAC) for print and the A-to-Z list for electronic holdings. Deciding whether or not to include print holdings in an A-to-Z journal list is important for a library because such a decision raises issues about the OPAC as a primary access point for journals. Accurate print and/or electronic holdings are critical for a good A-to-Z journal list, so when choosing one's first A-to-Z listing service or migrating to a new A-to-Z service, it is difficult to overstate the value of the global knowledgebase. How well a company maintains this knowledgebase and how often it is updated is critical to providing accurate information to the patron.

While many libraries have decided to outsource the maintenance of the A-to-Z lists to companies such as Serials Solutions, some libraries continue to maintain their own e-journal lists. For example, the University of North Carolina at Greensboro (UNCG) maintains its own knowledgebase and has developed its own link resolver, Journal Finder. Similarly, Simon Frasier University (SFU) developed its own journal listing solution and link resolver product, initially working with the Jointly Administered Knowledge Environment (JAKE) system from Yale University. SFU expanded on its initial experience with JAKE, and SFU staff subsequently developed CUFTS and the CUFTS knowledgebase. Finally, OhioLINK also developed its own solution, OLinks, which includes both the A-to-Z journal list and a link resolver. More information about these homegrown initiatives is available at their respective Web sites or in the January/February 2006 issue of *Library Technology Reports.*[3]

THE OPENURL AND LINK RESOLVERS

It is impractical to discuss A-to-Z listing services without concurrently addressing link resolver technology. While the A-to-Z journal list represented an important step forward by allowing a patron to easily discover whether a library has access to a particular journal electronically, link resolvers take the critical next step by offering connections among a library's collection of e-resources. Carol Tenopir notes, "linking to full text through link resolver technology and the OpenURL standard has made e-journals a cornerstone of library collections. Users expect that full text will always be a click or two away and it brings the library catalog, indexing and abstracting databases, and full text into an integrated system."[4]

A-to-Z journal lists were a boon for known-item searching, meaning they allowed a patron to quickly find and link into an issue of an e-journal. What of the patron already searching in an aggregated database or browsing the reference list of an article at the publisher's Web site? How can that patron easily link from a citation in one resource to the full text housed at another resource? Moreover, how can that patron link into the copy of the full text to which he or she has the rights to access? As evidenced by the *Journal of Black Studies,* any given journal can be available from a multitude of providers, and rarely does a library have access to all iterations of a journal. Thus, any linking framework needs to both link to disparate resources and also point patrons to the copy of a given e-resource appropriate for them, usually the copy of a given e-resource purchased for them by their local library. In short, a workable linking framework should take a user's context into consideration. As Grogg explains, "context-sensitive linking is just as it implies: it takes the user's context, usually meaning his or her affiliations but possibly also the user's intent for desired information objects, into account; therefore, ideally, context-sensitive linking only offers links to content the user should be able to access (i.e., appropriate copy)."[5]

More explicitly, two essential elements must be in place for a context-sensitive linking framework to function: (1) localized control (often via the knowledgebase), and (2) standardized transport of metadata, specifically the metadata which describes the users' desired information object. The OpenURL framework has both these elements. To satisfy the localized control requirement, "the library configures its local link resolver to match its holdings (print and electronic), and it defines what other links to applicable services it wants users to see."[6] This configuration of holdings is often one and the same for the A-to-Z list and the link resolver, meaning that one knowledgebase underlies both tools. Hence, libraries often chose to use the same vendor for their A-to-Z list and link resolver in order to have one data-entry point for staff. The second requirement, standardized transmission of metadata, is accomplished by virtue of the OpenURL's status as an ANSI/NISO standard, specifically Z39.88-2004.[7] The OpenURL is not the only option for context-sensitive linking, but it is one of the most widely used and the framework upon which many link resolvers are built.

OpenURL v. 0.1 and 1.0

Before dipping too quickly into a discussion of the functions of a link resolver, however, it is important to review the technology behind the OpenURL and its emergence as an ANSI/NISO standard. An Open-URL, unlike a traditional URL, does not point to a static address indicating the location of one copy of an object; "instead, the OpenURL contains metadata identifying the desired object, much like a MARC record identifies the item itself, not a specific copy of the item."[8] In addition to containing metadata about a given information object, the OpenURL also includes information about the user's context, usually his or her institutional affiliations. Moreover, the OpenURL is a dynamic linking technology, which means that the OpenURL itself is populated with metadata at the moment of clicking. Generally speaking, a user clicks on a link in an OpenURL-aware source, the OpenURL recognizes the user, and the OpenURL is sent to that user's local link resolver, which often contains the same localized knowledgebase that underlies the A-to-Z list. The metadata in the OpenURL is compared with the localized knowledgebase, and if a match exists, links to appropriate information objects are presented back to the user on a menu of services.

The OpenURL was initially in use as OpenURL v. 0.1, and in 2004, OpenURL v.1.0 emerged as ANSI/NISO standard Z39.88-2004. OpenURL v. 0.1 and its implementation via link resolvers was functional, and it addressed a real need: a context-sensitive solution that paved link paths from traditional bibliographic information to corresponding full text. As the technology matured, libraries began exploring what other information objects a user might want, thus moving beyond the traditional citation to full-text model. With the release of OpenURL v. 1.0, users see a more robust technology that is extensible enough to go beyond bibliographic to full-text linking and beyond scholarly communities as a whole.

The Digital Object Identifier and CrossRef

Just as it is difficult to discuss A-to-Z lists without simultaneously addressing link resolvers, it is impossible to discuss linking in the current scholarly realm without mentioning the Digital Object Identifier (DOI) and CrossRef. The OpenURL framework includes the standardized transmission of metadata about information objects. "Metadata can be familiar bibliographic information (author, title, journal title, or ISSN) or it can be the DOI, which is a persistent, unique identifier."[9] DOIs are administered by the International DOI Foundation and are registered by DOI Registration Agencies. CrossRef, a service of the Publishers International Linking Association, was the first and remains the largest DOI Registration Agency; it counts among its participants more than 1,600 publishers and societies.

The beauty of the DOI is that it is a persistent, unique identifier that does not change and can be associated with multiple instances of an article (at the publisher's site). Publishers "assign DOIs to each information object they publish and deposit these DOIs and the corresponding URLs in the CrossRef database."[10] CrossRef, then, works as a digital switchboard connecting DOIs with corresponding URLs. CrossRef also integrates with the OpenURL framework, and in the optimum scenario, the user is unaware of the complexity of the link exchanges occurring behind the scenes. The story does not end there, however.

As Grogg and Ashmore explain, "CrossRef does not limit itself to OpenURL and DOI linking." For example, "CrossRef has explored other ways in which it can facilitate scholarly research through expanded services such as CrossRefWeb Services, CrossRef Search, multiple resolution, free OpenURL and DOI resolvers, forward citation linking, and more."[11] Most recently, CrossRef announced its freely available Simple-Text Query Service to facilitate DOI lookup for researchers and publishers. This new service allows anyone "to retrieve DOIs for journal articles, books, and chapters by simply cutting and pasting the reference list" into a box made available on the CrossRef Web site (http://www.crossref.org/freeTextQuery/).

It is critical to note that DOIs link to the full text of information objects housed at publishers' Web sites. If full text is available elsewhere—for example, in an aggregator or in an Open Access (OA) source—then a link resolver becomes necessary, which, depending on the quality of the metadata and the knowledgebase, can present any number of full-text options, not just the publisher's full text. CrossRef remains an exemplar of what kinds of linking solutions are possible when publishers work with one another for a common goal: to facilitate an easier and more streamlined research process for users. Nevertheless, it is important to remember that CrossRef represents publishers, and its development efforts focus on increasing the use of the DOI.

Link Resolver Options

The behind-the-scenes technology of the OpenURL framework, CrossRef, and the DOI makes link resolvers possible. The link resolver itself relies on two critical components: the quality of the knowledgebase and the quality of its linking engine. A library deciding to purchase a commercially available link resolver, adapts an open source link resolver, or creates homegrown alternative needs to consider both of these components. Commercially available link resolver options include the following: LinkSource from EBSCO; SFX from Ex Libris; OL2 from Fretwell-Downing; LinkSolver from Ovid Technologies, Inc.; ArticleLinker from Serials Solutions/ProQuest; and TOUR Full Text Resolver from TDNet. Other possibilities include link resolvers available from Endeavor Information Systems, Inc., Innovative Interfaces, Inc., and Geac Library Solutions. As mentioned earlier, several universities, such as the (UNCG) and Simon Frasier University as well as some consortia, such as OhioLINK, have created homegrown solutions.

Beyond Traditional Linking

Context sensitive article linking has truly come into its own in a very short time. Perhaps the best evidence of this is the partnership between Google and a wide variety of linking vendors and providers to enhance the results of the Google Scholar search engine. Using OpenURL technology and capitalizing on partnerships with link resolver vendors, Google Scholar allows libraries who have a current link resolver to point patrons to library subscriptions from Google Scholar results. More information about what Google calls the "Library Links Program" is available at http://scholar.google.com/intl/en/scholar/libraries.html. The interest of such a prominent name in Web searching in what is essentially a library-born technology shows a move in the right direction for attracting users and providing them with preferred services. The importance of Google's acceptance and use of the OpenURL can not be overstated. Oren Beit-Arie, one of the original co-authors of the OpenURL v. 0.1, emphasized that "the acceptance of the OpenURL standard in a non-library source is a huge leap forward . . . the hope is that Google's acceptance will lead to further non-library acceptance of, and innovations with, the OpenURL."[12] With the April 2006 release of Windows Live Academic, Microsoft allowed OpenURL linking in its academic search engine (http://academic.live.com/). Other examples of new and innovative uses of the OpenURL include COinS (http://ocoins.info/), which, put simply, is a way to exploit the use of latent OpenURLs in Web pages. Finally, the acquisition of Openly Informatics, Inc.—"a pioneer in linking and the development of the OpenURL"[13]—by Online Computer Library Center (OCLC) demonstrates a commitment from OCLC to linking. Tenopir notes, "This partnership has the potential to be very fruitful, especially considering several other OCLC projects, such as the OCLC OpenURL Resolver Registry, OpenWorldCat, and eSerials Holdings."[14]

The importance of linking, particularly from a user standpoint, must not be underestimated. User expectations have been forever changed by Internet searching. As Grogg states, "It is, fundamentally, the librarian's job to facilitate the functionality that links these logically related resources. A user should not have to leave one system and essentially re-create his or her search in another to find the full-text gold at the end of the rainbow."[15] Beyond the simple linking relationship between a citation and a full-text article, linking represents a much larger opportunity for scholarly discourse. The original creators of the OpenURL, Van de Sompel and Hochstenbach, describe the full value of linking as "ways to create added-value by linking related information entities, as such presenting the information within a broader context estimated to be relevant to the users of the information."[16]

FEDERATED SEARCHING

Whereas A-to-Z lists facilitated known-item searching and the OpenURL—via link resolver implementation—provided critical links amongst resources, federated searching addresses a different need: the ability to search across many e-resources. Federated searching represents not only the opportunity to make library search services resemble search engine competitors, but it also gives a library the opportunity to guide users to resources they most likely would have shunned. Karen Calhoun's report for the Library of Congress—"The Changing Nature of the Catalog and its Integration with Other Discovery Tools"[17]—makes much of the declining use of OPAC searching, illuminating users' unlikelihood of searching for information in the catalog. The OPAC appears to appeal to users only for known-item searches and not broader topic searching. Federated searching is one possible way to boost the OPAC's topic searching capabilities. By overlaying OPAC interfaces with a more familiar, powerful, and inviting federated search interface, users can pull up catalog records. Just as the Z39.50 standard made searching more than one catalog possible for users, coupled with XML technology, Z39.50 continues to broaden the definition of database searching for library users.

Federated searching, while still a fairly young technology, has demonstrated great potential to empower users. In an October 2004 *Information Today* column, Peter Jasco notes:

Federated searching consists of transforming a query and broadcasting it to a group of disparate databases with the appropriate syntax, merging the results collected from the databases, presenting them in a succinct and unified format with minimal duplication, and allowing the library patron to sort the merged result set by various criteria.[18]

This is not to say that this is federated searching as library patrons know it today. Issues like user authentication management, effective de-duplication, and flexibility in display are still being explored by vendors, but there is progress.

NISO's Metasearch Initiative (http://www.niso.org/committees/ MS_initiative.html) is bringing together vendors, libraries, and other stakeholders to address the issues mentioned in previous text. This Metasearch Initiative is also addressing copyright concerns, search protocol standardization, and common descriptors and tags for content.[19] Mark H. Needleman, in a 2006 update on the Metasearch Initiative's activities, summarizes the challenges and potential rewards of metasearch:

Lack of standard mechanisms for authentication, search and retrieval, and metasearching, while currently possible and being implemented and used, puts a strain on both the metasearch systems and the content and data providers. . . . If the standards and protocols discussed above [in his article] take hold and are implemented, metasearching will become more efficient, and metasearch systems and content providers will be able to interoperate in a manner that is more efficient for both of them.[20]

In addition to the issues tackled by NISO, librarians are well aware of other pertinent considerations. As Rachel Wadham puts it, "more is not necessarily better and without the right match of product and library we may find we are giving our patrons more access not necessarily better access."[21] Federated searching represents a direct response to users' behaviors and preferences. However, the role of the librarian in federated searching remains central. Through careful selection and grouping of library resources, libraries still provide value-added services—services that are not available via free Web searching. Oliver Pesch offers the following explanation of federated searching:

the technology that makes the process [of federated searching] seem simple is actually rather complex. Unlike Google and Yahoo, federated search engines do not actually "index" the Web sites they search; instead, they use special software that performs a search on the actual Web site in real time. This software, known as a translator or connector, is customized for each Web site, and often needs to be customized for each database.[22]

Pesch aptly notes that the technology behind federated searching is anything but simple, and he goes on to comment about the time, energy, and planning involved in implementing a federated search system. While vendors may promise quick implementations ranging anywhere from six weeks to three months, libraries have discovered that implementation often takes much longer.

Librarians must decide whether the payoff is worth the effort. In a comparison between Google and two of the most popular federated search tools, MetaLib and WebFeat, Xiaotian Chen notes that federated searching does lead users to databases they might otherwise not have chosen, but also adds several caveats:

But in no way can the federated search compete with Google in Google's strengths: speed, simplicity, ease of use, and convenience. Nor can the federated search truly serve as one-stop shopping for all library databases as people hoped, because some databases cannot be searched by the federated search for various reasons.[23]

Indeed, librarians must be realistic about the current capabilities of federated searching; moreover, librarians must first decide the primary purpose of federated searching in the local environment before jumping on the next metasearch bandwagon.

In many ways, federated searching has entered its adolescence. In 1998, WebFeat emerged as first product to offer federated searching as we know it today.[24] Since 1998, the marketplace has expanded to include products such as MetaLib from Ex Libris; Central Search from Serials Solutions; Discovery: Finder (formerly ENCompass) from Endeavor Information Systems, Inc.; Ovid SearchSolver from Ovid Technologies, Inc.; and most recently, Research Pro from Innovative Interfaces, Inc. Whereas A-to-Z lists and link resolving have been widely implemented and accepted (in North America) as reliable e-journal management tools, federated searching is still experiencing growing pains and its future remains to be seen.

THE FUTURE OF E-JOURNAL MANAGEMENT TOOLS

A-to-Z journal lists, link resolvers, and federated search products are certainly not the only ERM tools on the market. As mentioned earlier, MARC record services provide a critical service for libraries by offering periodically updated MARC records for electronic subscriptions. Serial vendors and subscription agents continue to reevaluate their roles in the process; for example, the enhanced version of EBSCO*host* Electronic Journals Service (EJS) promises to offer "extensive features that help with e-journal management tasks such as: tracking the registration status of e-journals, authentication assistance to facilitate both on-campus and remote access to e-journal content, automatic management of e-journal URLs and much more" (http://www.ebsco.com/home/ejournals/default.asp).[25] Other products such as SwetsWise from Swets Information Services promise similar services. In addition, ERMS—the need and specifications for which are outlined by the Digital Library Federation's Electronic Resource Management Initiative (http://www.diglib.org/standards/dlf-erm02.htm)—are another spoke in the e-journal management wheel.

While each tool addresses a different need, they all have a common goal: helping librarians manage and provide access to a chaotic and exponentially increasing c-rcsource collection. Libraries are experiencing a profound paradigm shift, and the space between print as primary mode of access and electronic as primary mode of access is unstable, confusing, and disorganized—an uncomfortable situation for librarians. Print is no longer king for most library patrons, yet libraries have spent centuries honing and developing ways to process and make available print resources. Instead of relying on these tried and true technical processes or trying to adapt them to an electronic world, librarians must explore the best ways in which to implement new and innovative e-journal management tools.

NOTES

1. Christine L. Ferguson, Maria D. Collins, and Jill E. Grogg, "Finding the Perfect E-Journal Access Solution . . . the Hard Way," *Technical Services Quarterly* 23, no. 4 (2006): 28.

2. Serials Solutions, "Serials Solutions—Home" (2000-2006), http://www.serialssolutions.com/home.asp (accessed December 8, 2006).

3. Jill E. Grogg, *Linking and the OpenURL: Library Technology Reports* 42, no. 1 (January/February 2006): 24-30.

4. Carol Tenopir, "Thinking About Linking," *Library Journal* 131, no. 12 (July 1, 2006): 29.

5. Grogg, *Library Technology Reports,* 14.

6. Tenopir, "Thinking About Linking," 29.

7. National Information Standards Organization, "Z39.88-2004: The OpenURL Framework for Context-Sensitive Services," (April 15, 2005), http://www.niso.org/standards/standard_detail.cfm?std_id=783 (accessed December 8, 2006).

8. Grogg, *Library Technology Reports,* 15.

9. Tenopir, "Thinking About Linking," 29.

10. Grogg, *Library Technology Reports,* 17.

11. Jill E. Grogg and Beth Ashmore, "CrossRef at the CrossRoads," *Searcher: The Magazine for Database Professionals* 14, no. 8 (September 2006): 33.

12. Grogg, *Library Technology Reports,* 39.

13. Tenopir, "Thinking About Linking," 29.

14. Ibid.

15. Grogg, *Library Technology Reports,* 5.

16. Herbert Van de Sompel and Patrick Hochstenbach, "Reference Linking in a Hybrid Library Environment, Part 1: Framework for Linking," *D-Lib Magazine* 5, no. 4 (April 1999): http://www.dlib.org/dlib/april99/van_de_sompel/04van_de_sompel-pt1.html (accessed December 8, 2006).

17. Karen Calhoun, "The Changing Nature of the Catalog and Its Integration with Other Discovery Tools," Final Report (March 17, 2006), http://www.loc.gov/catdir/calhoun-report-final.pdf (accessed December 8, 2006).

18. Péter Jacsó, "Thoughts About Federated Searching," *Information Today* 21, no. 9 (October 2004): 17.

19. Donna Fryer, "Federated Search Engines," *Online Magazine* 28, no. 2 (March/April 2005): 18.

20. Mark H. Needleman, "An Update on the NISO Metasearch Activity," Serials Review 32 (2006): 144-145.

21. Rachel Wadham, "Federated Searching," *Library Mosaics* 15, no. 1 (January/February 2004): 20.

22. Oliver Pesch, "Re-inventing Federated Searching," *Serials Review* 32 (2006): 183.

23. Xiaotian Chen, "MetaLib, WebFeat, and Google: The strengths and weaknesses of federated search engines compared with Google," *Online Information Review* 30, no. 4 (2006): 426.

24. Pesch, "Re-inventing Federated Searching," 183.

25. EBSCO Information Services, "E-Resource Access and Management," 2006, http://www.ebsco.com/home/ejournals/default.asp (accessed December 8, 2006).

Chapter 13

Creating an E-Resource Infrastructure: A Case Study of Strategies At Seven Academic Libraries

Glen Wiley

INTRODUCTION

Since the mid-1990s, the number of e-resources available to library users has grown dramatically. Libraries now provide users with access to myriad e-journals, e-books, digital images, electronic indexes and abstracts, and many other e-resources. As libraries make this transition in their collections, successfully managing and providing users with access to e-resources has become a growing challenge. While libraries initially relied upon such tools as Integrated Library Systems (ILS), hand-coded HTML pages, and Access databases publishable to Web pages, there has been an increasing realization that the effective management and access to e-resources necessitates the development and maintenance of an infrastructure of integrated tools. To meet this need, a variety of new tools have been unveiled by vendors and developed by libraries. As these tools continue to evolve with new services and new systems, there are a growing number of paths that a library can take in order to create an e-resource infrastructure. Selecting which path to take in its development of an e-resource infrastructure requires that a library assess the unique combination of e-resource access and management tools that will meet the varying needs of both users and the library itself.

This chapter discusses the overarching criteria and fundamental reasons that lead a library to develop a particular combination of tools to manage and provide access to its e-resources. To explore this topic, the chapter specifically focuses on seven academic libraries' strategies for selecting, implementing, and utilizing technology to create an e-resource infrastructure. It begins by investigating the specific infrastructures created for e-resource management at the seven libraries studied, focusing primarily on the suites of tools that these libraries use. Next, the chapter discusses the rationales for the creation of each infrastructure and draws connections between these rationales and factors such as library size, organizational structure, and user needs. Finally, the chapter evaluates the relative strengths and weaknesses of the libraries' strategies for e-resource access and management and discusses lessons that can be learned from an analysis of these libraries' e-resources infrastructures.

METHODOLOGY AND GOALS
OF THE DATA COLLECTION

To explore the challenges libraries face in developing an e-resource infrastructure, data was gathered from seven academic libraries using surveys, questionnaires, and phone interviews. Since many of these libraries did not want to be identified with their particular views, each of the surveyed libraries is coded with a number rather than identified by name. While the means of the data collection is not scientifically rigorous enough to warrant prescriptive recommendations or broad-reaching statements about all libraries, it does reveal the factors that shaped the e-resource infrastructures of seven academic libraries of varying sizes, locations, and collections. In doing so, the chapter's analysis aims to both capture the realities of combining e-resource tools together and examine the variable nature of e-resources infrastructure as a whole. Ultimately, the chapter should reveal larger concerns related to these tools and provide new insights that can be applied to the development of any library's e-resource infrastructure.

OVERVIEW OF THE CASE LIBRARIES
AND THEIR E-RESOURCE INFRASTRUCTURES

The seven academic libraries examined in this case study utilize a variety of different products and services from a variety of different sources. Table 13.1 provides an overview of the libraries studied and a summary of the tools that these libraries use. From this table, it is possible to make some basic observations about the case libraries and their tools for front-end and back-end e-resource management. First, most libraries are choosing to transition from homegrown products to commercial products. Indeed, most of the libraries surveyed have already or are currently in the process of moving away from in-house products. The case libraries appear to be finding that the total cost and time spent on those in-house products are increasingly outweighed by commercial products, which show a greater potential to add more value to managing and displaying e-resources. A second observation to be garnered from Table 13.1 is that commercial e-resource tools are generally suitable for more than one type of library. Indeed, this table includes a number of instances in which several different types of academic libraries with a variety of e-journal collection sizes use the same link resolver, subscription management system, or electronic resource management (ERM) system. Third, while the case libraries have chosen and implemented an e-journal vendor, an ILS system, an open URL link resolver, and an A-Z e-journal list, these libraries have been slower to implement a federated search product and ERM system. The likely reason for this fact is rooted in the cost, development, and maturity of commercial ERM systems and federated search products.

INFRASTRUCTURE RATIONALES

Over a period of many years, libraries have fine-tuned their processes and tools for successfully serving their users in the print environment. The electronic environment, however, is still unmapped, and there are no clear guidelines for success. Although all libraries attempt to implement the tools that will most effectively serve their needs for the access and management of e-resources, the factors that determine what tools to purchase and implement are unique to each institution. Ultimately, several primary factors can be identified that drive a library's e-resource decisions and contribute to the end combination of tools that the library uses. Among the factors shaping a library's e-resource infrastructure are the selectors of tools within the

TABLE 13.1. Overview of the Case Libraries and Their E-Resource Infrastructures

Library Number	Type of Academic Library	E-Resources Collection	Subscription Management System	ILS System	Open URL Link Resolver	E-Journals A-Z List	Electronic Resources Management System	MARC Record Service	Federated Search Product	Other Tools For E-Resources
1	Large Southeastern public university library	31,921 e-journals, 193,447 e-books, and over 200 databases	EBSCOnet Subscription management system	Endeavor's Voyager ILS	Ex Libris' SFX link server	Homegrown A-Z lists for all e-resources created by daily lists from Voyager ILS	Ex Libris' Verde ERM System.	MARCIVE MARC record service and several other vendor record sets, but no universal MARC record service for e-journals	None	None
2	Large Northeastern public university library	35,320 e-journals, over 350 databases, and 9,379 e-books	EBSCOnet Subscription management system	Endeavor's Voyager ILS	Ex Libris' SFX link server	Serials Solutions Access and Management Suite	An in-house ERM system, but they hope to purchase a commercial system in the near future	Serial Solutions' Full MARC record service	None	None
3	Small, Midwestern private college's library	17,667 e-journals, 104 databases, and no e-books	EBSCOnet Subscription management system	Endeavor's Voyager ILS	Ex Libris' SFX link server	A-Z list derived from Ex Libris' SFX knowledge base	None	Ex Libris' MarcIt! MARC record services	WebFeat federated search engine	None

4	Medium-sized, West Coast public university library	Over 25,800 e-journals, e-books, and databases	EBSCOnet Subscription management system	Innovative Interface's Millennium ILS	Ex Libris' SFX link server	A-Z list derived from Ex Libris SFX knowledge base	Innovative Interfaces' ERM	None	Ex Libris' Metalib metasearch tool	None
5	Small, Southeastern private college's library	Over 23,000 e-journals, 10,000 e-books, and 120 databases	EBSCOnet Subscription management system	Innovative Interfaces' Millennium ILS	Serials Solutions' Article Linker	Serials Solutions' AMS service	Innovative Interface's ERM	Serials Solution's MARC record services	Meta-search tools are currently being evaluated	In-house developed Database Finder
6	Large, Midwestern private university library	14,900 e-journals, 11,939 e-books, and 209 databases	EBSCOnet Subscription management system	Endeavor's Voyager ILS	Ex Libris' SFX link server	A-Z list derived from Ex Libris SFX knowledge base	No ERM, but they will be building a home grown system based on the DLF ERM Initiative	MARCIVE MARC record service and will contract for specific MARC cataloging projects as needed	None	Library Management system
7	Medium-sized, private, Northeastern university library	Over 43,000 e-journals, 10,000 e-books, and 170 databases	EBSCOnet Subscription management system	Innovative Interfaces' Millennium ILS	Ex Libris' SFX link server	EBSCO A-Z service	Innovative Interfaces' ERM	MARCIVE MARC record service and other vendor-supplied record sets	None	None

library, the financial resources available, the cooperation among libraries, the library staff's knowledge, and the order in which tools are acquired.

The Selectors of E-Resource Tools

The case libraries varying organizational structures and staff sizes were among the primary factors determining who is responsible for selecting their e-resource tools. The majority of these libraries have created specially formed committees to analyze several different products and then decide which e-resource management tools to implement. The committee members generally represented a variety of library departments. For example, in library number three, a library committee comprised of two electronic resources librarians, the head of Technical Services, a technical services staff member, and a reference staff member selected all of the library's e-resource management tools.

In several of the other libraries surveyed, however, sole decision makers can also determine whether to acquire a tool for e-resource management and access. In library number five, for example, the electronic services librarian recommends to the library's director which e-resource management tools should be implemented. Likewise, library number four notes that its assistant director for systems and planning made the decision to purchase Ex Libris' SFX and MetaLib products. Library number one takes more of a middle-ground approach in which it selects e-resource tools in two ways. In some cases, products are acquired based on the recommendations of the coordinator of technology services. However, some higher-impact tools, like an ERM system and metasearch product, are selected based on the recommendations made by a committee formed for that special purpose.

Financial Constraints

As libraries make the transition to e-resources, they are also making a transition in their material expenditures. For example, the Association of Research Libraries (ARL) has reported an annual increase in e-resource expenditures that averages about 28.77 percent from 1993 to 2004.[1] Not surprisingly, given this dramatic increase in e-resource expenditures, the funding available for and cost of e-resource management and access tools were top considerations in the seven case libraries' efforts to develop e-resource infrastructures. The case libraries commented on the growing strain required to maintain funding for all of their e-resource tools as these tools increase in both quantity and cost. For example, library number three notes that, while it considers a wide range of tools for e-resources irrespective of cost, its ultimate decisions are deeply rooted in its limited budget.

Constraints on funding for technology can influence whether a library buys a commercial product, installs an open-source tool, or creates a homegrown device. Four out of the seven case libraries indicated that they had developed in-house tools for e-resource management and access that were eventually replaced by commercial products. These in-house products were usually e-resource databases using software (such as MySQL, FileMaker Pro, or Microsoft Access) and scripting languages (such as PHP and Cold Fusion) in order to dynamically display information from the database on a library Web site. Although these in-house products offered locally customized solutions to the libraries' e-resource management and access needs, they were also time-consuming to develop and maintain. As vendors' commercial products have been enhanced and as libraries' limitations on financial and human resources have become more marked, the case libraries indicate that they readily prefer available commercial products. An exception to this is library number two. While this library indicates that it tries to avoid homegrown solutions because of the cost and time involved, it does favor open source over proprietary tools whenever open-source tools are readily available and can be easily implemented.

Cooperation Among Libraries

Whether taking the form of loosely formed consortia or highly structured organizational partnerships, cooperation among libraries can have a decisive impact on the choice of tools for e-resource management and access. Four out of the seven case libraries note that they do indeed coordinate the acquisition of tools for e-resources within a consortium. E-resource committees are the most common means to coordinate such acquisitions. The degree of coordination and partnership varied among the case libraries. In one instance, coordination was limited to using the same union catalog and link resolver. In another instance, coordination with partner libraries included sharing training sessions and ERM resource records for the shared resources within the same commercial tool.

The Library Staff's Knowledge

All seven of the case libraries have made staffing changes as a result of their shift from print to e-resources. Their new and evolved staff positions cover everything from collection management to technical services and Web services. These positions require staff members with new skill sets that will allow for the introduction and maintenance of new technologies. Accordingly, all of the case libraries indicated that investing the time to educate staff members about new e-resource tools is invaluable. Indeed, staff members' general knowledge of a product and its relationship to their specific job responsibilities can help determine the viability of a tool being implemented within a library. Product demonstrations, conference presentations, published reviews, and listserv discussions were all cited by the case libraries as ways of educating library staff members about new e-resource tools. Library number seven provided a specific example of how staff knowledge was utilized in its development of an e-resource infrastructure. This library needed to evaluate the functionality of Innovative's ERM system in order to determine the feasibility of replacing the Access database that it had previously used to track e-resource licensing and access terms. To carry out this evaluation, staff members were asked to list and categorize the functions of each system; this exercise provided the library with valuable insights about the impact of Innovative's ERM system on its current management of e-resources.

When evaluating a new e-resource tool, the case libraries emphasized the importance of tapping staff members' previous experiences with the tool's provider. Indeed, a high level of satisfaction among library personnel with a particular vendor provides evidence that the vendor will offer an equally strong level of reliability and customer support for the tool being evaluated. One example of how staff knowledge impacts a library's e-resource infrastructure is described by library number two. This library switched from Link Finder Plus to SFX because of the experience their new electronic resources librarian had with SFX from a previous job; indeed, the librarian's favorable opinion of Ex Libris' SFX product and its customer service ultimately led her new library to switch link resolver software.

The degree of staff members' programming and systems-related knowledge can also impact the selection of e-resource tools. For example, due to its staff members lack of programming knowledge and inability to provide in-house systems support, library number one noted that it wants to implement mature products that have already proven to be user-friendly. Indeed, this library has been looking at ERM systems for over a year now, but has delayed a decision because none seemed to be mature enough until recently, when the library expressed interest in implementing Ex Libris' ERM software, Verde. The library hopes that Ex Libris can provide the same high level of systems support for its ERM system that the company currently provides for the library's link resolver, SFX.

The Order of Acquisition

For most of the libraries in this study, e-resource tools were acquired in roughly the same order. Although it is clear that many factors influenced this order of acquisition, among the most important factors were cost, the maturity of commercial products, the time available for staff to focus on implementing new tools, user demands, and the reorganization of departments. Among the earliest tools to be acquired by the case libraries were proxy software, A-Z e-journals lists, and link resolvers. As was noted in an earlier section, the two most recent tools to be acquired by the case libraries are federated search engines and ERM systems. Due to technical barriers and the need to develop standards, these tools are relatively new additions to the e-resource marketplace that can often bear a significant price tag. Regarding federated searching (also commonly referred to as cross searching and metasearching), three out of seven case libraries are currently in the process of evaluating a federated search engine for their databases while two (libraries number three and four) already have federated search products in place. Regarding ERM systems, five out of the seven case libraries have implemented products while two (libraries number three and six) have not.

STRENGTHS, WEAKNESSES, AND LESSONS LEARNED

All seven of the case libraries discussed the strengths and weaknesses of their e-resource infrastructures. Beyond the vendor-specific and tool-specific commentary, the case libraries expressed common realizations regarding their strategies for building a customized e-resource infrastructure.

One of the case libraries' realizations was that licensing a suite of e-resource tools from a company like Serials Solutions, EBSCO, or Ex Libris could strengthen their infrastructures. While the majority of libraries did not stay with any one commercial company's products, in retrospect they did recognize the benefits of this approach. Among the advantages of purchasing a suite of tools from one company is that the user and administrative interfaces are consistent across tools, the library only needs to maintain one e-resource knowledgebase, and the library has the opportunity to forge a close partnership with a single vendor. Without a suite of tools from a single vendor, many of the case libraries have been forced to develop workflows in which they must individually update such tools as a link resolver service, A-Z list of e-journals, and catalog record in order to coordinate the same data and accessibility in different tools. This concern for one data source is especially timely given the case libraries' acquisitions of ERM systems. Indeed, because of the desire to manage all e-resources from a central location, the integration of an ERM system can call into question many of a library's previous decisions regarding the implementation of e-resource tools. For example, library number four would like to use its SFX knowledgebase to load information into Innovative's ERM system. However, it is not possible to easily get certain types of information out of the SFX knowledgebase without utilizing scripting and other means to manipulate data, which is a barrier that limits the library's integration of content within its e-resource infrastructure.

Along with the advantages discussed in previous text of the single vendor approach for e-resource management and access, there are also drawbacks. One drawback is that the vendor's knowledgebase may not entirely reflect the library's holdings due to unique contracts, packages, and titles. Indeed, some e-resource tools have limited knowledgebases or lack the level of detail needed to manage and display correct holdings and notes. Another drawback to a single-vendor approach is that it prevents a library from developing an e-resource infrastructure that takes full advantage of the relative strengths and weaknesses of different vendors' products. For example, a vendor may have an excellent link resolver, metasearch product, and MARC record service, but an ERM system with capabilities that are inferior to its competitors. If a library goes with a single vendor for all of its e-resource access and management tools, it would not have the flexibility to select an ERM system that more effectively meets its needs.

A second realization that the seven case libraries commented on was having one e-resource tool provide multiple services. For example, many libraries have used data reports in spreadsheet-friendly formats from e-resource tools for many purposes. Some libraries, including library number seven, use their A-Z title management list data to export their e-journal information into their catalog instead of purchasing a MARC record service. The case libraries also noted that the spreadsheet data can be easily used to create title lists and explore collection development activities like e-journal overlaps analysis.

A third realization discussed by the case libraries was the need to be creative and share information about e-resource tools with staff members throughout both technical services and public services. The involvement of diverse staff members allows for a better and more multifaceted understanding of a tool's capabilities and limitations. For example, when implementing a link resolver, library number seven assigned responsibility to both technical services staff members and reference librarians. The library's choice to involve multiple personnel in the implementation process resulted in a fast implementation in which no single librarian or department was overwhelmed by the new tool.

One last realization by the libraries surveyed was the need to carefully and deliberately select e-resource tools. Regardless of pressure from users, the selection process should not be one that is rushed or carried out impulsively. Library number four, for example, made quick decisions on several e-resource tools that it is now reconsidering. If this library had waited, it notes that it would probably have purchased tools with one vendor instead of distributing these products across vendors. The varying data sets have different interfaces, varying data structures, and many other disparate standards, which make the tools harder to use and integrate. Indeed, in evaluating e-resource tools, libraries should think about the implications of local conditions and the long-term sustainability of their e-resource infrastructure.

CONCLUSION

Numerous articles, presentations, and book chapters have been written to address the nuances of individual e-resource tools. However, the challenges of integrating these tools into an effective infrastructure are topics that remain largely unexplored. This case study of seven different academic libraries gives a snapshot of the variety and complexity of challenges libraries have struggled with in designing e-resource infrastructures. The key to effectively overcoming these challenges is for a library to assess its environment, study the tools available, survey local needs as well as limitations, and dedicate funding that will result in a successful e-resource infrastructure.

NOTE

1. ARL Statistics, Articles, Introductions, Graphs, and Tables for 2003-2004, "Graph 6: Yearly Increases in Average: Electronic Resources vs. Total Materials Expenditures, 1993-2004,"Association of Research Libraries, http://www.arl.org/stats/arlstat/graphs/2004/ematbar.pdf (accessed January 11, 2007).

PART IV:
EVOLVING STRATEGIES
AND WORKFLOWS

Chapter 14

Analyzing Workflows and Realizing Efficiencies for Serials Processing

Elizabeth S. Burnette

INTRODUCTION

In the past decade, e-resources have become a significant portion of the serials processed in the technical service departments of libraries. Unfortunately, databases, e-journals, e-books, and other e-resources do not fit neatly into preexisting workflows designed to process print serials. As a result, workflow managers have had to make major changes to these preexisting workflows while developing new workflows to effectively process their libraries' evolving collections. In order to maintain efficiency in this highly dynamic environment, periodic analysis of serial workflows is essential. This chapter provides guidelines through which a manager of serials processing can carry out workflow analyses in order to succeed in an environment increasingly dominated by e-resources. In particular, the chapter outlines a plan for how workflows for processing serials in both print and electronic formats can be analyzed in order to uncover and then eliminate inefficiencies. The chapter begins by discussing a few of the most important issues impacting serial workflows and reviewing the basic acquisitions cycle and workflows for serials. Building on this information, the chapter provides managers with general guidelines for analysis, design, and planning of workflows. The chapter concludes with an exploration of specific examples for realizing efficiencies throughout technical service units with responsibilities for processing serials.

ARE SERIAL WORKFLOWS ALL THEY CAN BE?

Like businesses, libraries realize that remaining viable and competitive in today's information marketplace requires that they "understand and improve the structure and execution of [their] business processes."[1] To achieve this goal, library administrators are analyzing the state of libraries and asking timely questions regarding users' expectations, libraries' relative strengths and weaknesses, and the trends that will shape the information marketplace in the future. Such a critical analysis of library resources and services inevitably highlights the importance of workflow efficiency as a means to compete for users and meet their evolving expectations; indeed, without efficient workflows, the quality of a library's resources and services will suffer and, as a result, users will turn to alternative sources to meet their information needs. It is for this reason that workflow planning and analysis are crucial components in a library's success. Achieving this success with respect to the library's access and management of serials is rooted in an understanding of the major issues that shape serial workflows.

Environment is among the most important factors impacting serial workflows. A workflow's environment consists of such components as a library's: (1) policies, (2) staff resources, (3) serials management tools, and (4) time allocated to achieve objectives. A change in any of these environmental factors is a sign that workflow redesign may be necessary. If not addressed, such a change can either lead to a breakdown of workflows or a missed opportunity for enhancing workflows. For example, when new IT tools change a library's environment, managers and staff members both have roles to play in order to effectively integrate the new tools into existing processes. Through planning or serendipity, staff may become aware of the potential uses for a new software functionality in serial workflows. Before it can be utilized in order to increase the efficiency of workflows, however, staff members need to fully understand this functionality, which can only happen with the support of their manager. Indeed, when managers properly equip staff members through training and development, staff can think creatively in order to enhance workflows.

Not only do changes in environmental factors indicate that a workflow review may be necessary, but bottlenecks and backlogs are also a sign that a workflow needs to be reviewed. Bottlenecks arise when the work going into a process exceeds the rate of production. Departments should identify bottlenecks to determine resource needs and resolve the production problem. Bottlenecks can be caused by sudden or steady changes in the resources available to perform a task. A backlog, in turn, is a visible sign that a bottleneck exists; in other words, it is an indication that the resources to carry out a particular process are insufficient or need to be redistributed. Managers must understand resource needs to devise a plan to prevent or eliminate backlogs. They may also need a plan to chip away at the backlog, borrowing staff from other units when possible. Backlogs are particularly dangerous in the e-resource realm because physical symptoms may not be visible until users' access is adversely impacted.

Just as important as recognizing the factors that impact workflows is understanding the relationship between these workflows and overall goals of the library. By drawing connections between the library's goals and the serials process, managers can help staff members develop an understanding of the "whys" of the work that they perform. In virtually any library, one of the core goals is to provide users with access to the information that they need. For the units responsible for carrying out serial workflows, this means that staff members should strive for efficiency, accuracy, and cost-effectiveness in all their tasks. Given such a goal, the objectives for serial workflows become the following: (1) to minimize the time needed to acquire serials, (2) to maximize all available staff and serials management tools, and (3) to establish procedures that insulate against attrition. Before discussing how workflows can be developed that meet these objectives, it is useful to first review the basic acquisitions cycle and workflows for serials.

Serial Workflows

Workflows for processing any serial resource can be organized into five distinct phases: selection, order, payment, access, and storage; the tasks within these phases are designed to achieve the following goals:

- identify the title;
- verify its bibliographic data;
- identify the price;
- assess availability;

- submit a requisition;
- submit a purchase order to the source;
- receive the title ordered or claim what is missing;
- receive an invoice;
- pay the invoice;
- store what is acquired;
- renew or cancel the title.

Although there are a number of variables that can complicate this process (such as the source and model through which the resource is acquired), this chapter limits its discussion of serial acquisitions to the process's most general and basic components, which are outlined in Figure 14.1.

Of course, the workflows needed to carry out the process outlined in Figure 14.1 differ for serials in print and electronic formats. As is stated in the 2004 report of the Digital Library Federation's Electronic Resources Management Initiative, "while there are some similarities

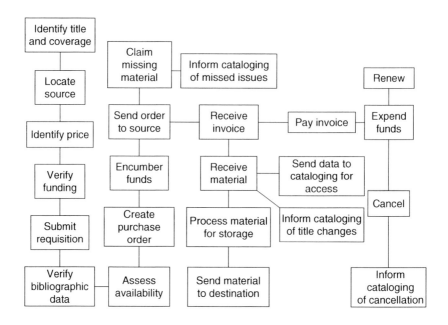

FIGURE 14.1. The Serial Acquisitions Process

between the acquisition and management processes for traditional physical library materials and those for electronic products, there are many issues and complexities unique to electronic products."[2] Figures 14.2 and 14.3, which provide samples of basic workflows for serials in print and electronic formats, illustrate the differences between print and electronic materials processing.

Although the workflows depicted in Figures 14.2 and 14.3 may appear cut-and-dried, they can also be intimidating. What some may find difficult about these workflows is that their designs have largely been shaped by changes in the publishing industry rather than by the decisions of the library. In other words, a change in format led to a change in the library's workflows, and the workflow design process occurred in reaction to outside forces at a time when staff expertise was low. The outcome of this is a sense among staff members that workflows are outside their library's control. The desired condition

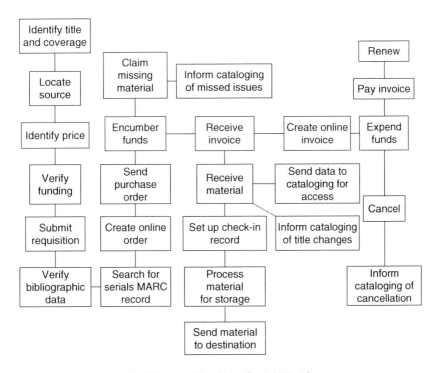

FIGURE 14.2. The Print Serial Workflow

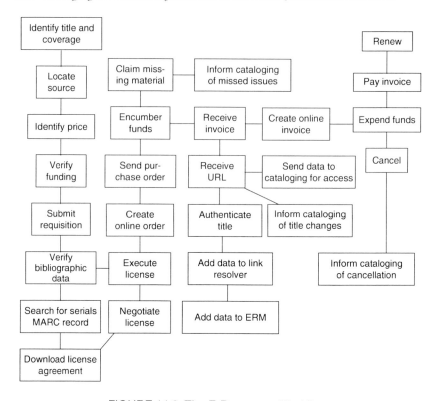

FIGURE 14.3. The E-Resources Workflow

for workflow design is proactive and methodical. Such a deliberate workflow design is better planned and understood before it is implemented, contributing to a sense of confidence among the staff members who carry the workflows out. As libraries have become more accustomed to processing resources in electronic formats, they now have the opportunity step back in order to proactively enhance these workflows through analysis and planning.

Analyzing Print Serial Workflows

Despite the numerous differences between workflows for serials in print and electronic formats, these workflows both require that staff members successfully fulfill the serial acquisitions cycle. Indeed, workflows for both formats must achieve the following objectives: preserving the audit trail, ensuring the renewal of one subscription per year, and tracking the receipt of what was purchased. Owing to this connection between workflows for serials in print and electronic formats, the development of workflows for e-resources provides managers with an opportunity to review print serial workflows in order to ensure that consistent standards prevail across all serial processes. An objective of print serial workflows review is the hunt for inefficiencies that, when found and corrected, will make available more resources that can be allocated toward acquiring access for e-resources; as Yue and Anderson state, "libraries have been reallocating staff or redefining positions to focus more staff on the challenging jobs of acquiring and maintaining electronic resources."[3]

A manager must utilize a variety of tools in order to conduct a thorough analysis of the print serial workflow. These tools include

1. staff experienced in the day-to-day functions of the unit;
2. documentation;
3. sample titles and issues;
4. sample records in the technology used to manage the process;
5. a flow chart (such as the one in Figure 14.2).

This workflow can be used to frame the entire process and highlight steps within the process. With these tools, a manager can develop a plan in order to

1. review and analyze existing print serial workflows;
2. educate staff on the process;
3. identify areas of concern;
4. collaborate on solutions;
5. communicate changes.

The steps in this plan, which should be adopted to accommodate the unique characteristics within an individual library, are as follows:

1. Review existing documentation, and revise it to reflect any edits that occurred since the last revision. During this step, poll staff members for edits if the process has not been revised recently or if recent changes in environmental factors have occurred.

2. Involve staff in the analysis. The staff members that perform the daily operations in a print serial workflow see both the emerging trends

and the issues that impact procedures. They can do much to contribute to the analysis and suggest ideas to increase efficiency.

3. Review the policies, guidelines, and philosophies that impact the direction of relevant units within the library. When doing so, be aware of new initiatives that administrators may have underway.

4. Discuss the existing workflows with staff members and look for tasks that have become obsolete. Walk through the process in order to identify exceptions and trends. Also, solicit ideas from staff for consideration during the revision phase of the workflows design.

5. Create a provisional revised workflow and distribute the draft (marked as such) to staff members. Provide specific deadlines, ask them to review, test, and respond to the draft. After receiving staff members' feedback, determine if the suggested revisions are viable and then edit and retest as needed. In addition, solicit feedback from other units and staff members impacted by the proposed changes. The objective here is to create a transparent environment in which the upcoming workflow changes will be received. When resistance, reluctance, or reservations to the change are valid, consider the issues and possibly a probationary implementation of the workflow, followed by a review of the experiment by all stakeholders, an assessment of the process, an analysis of any statistics collected, and follow-up discussion.

6. Look for links between the new ideas and current policies and guidelines. This process may initiate change and require buy-in from other managers and stakeholders. If so, prepare a written explanation of: (1) the workflow issue or problem, (2) the policy affected by the proposed workflow revision, (3) the justification for the change, and (4) the implications of not changing the workflow. Doing so will help managers relate the results of their decisions and any residual effects for future issues, planning, and policies.

7. Implement the revised workflow after it has been accepted. This entails distributing documentation concerning the workflow and establishing a start date. Set aside time to address staff members' questions during the training phase of the new workflow.

8. Set a date to meet with staff members about their experiences prior to any open meetings and keep administrators abreast of progress.

9. Cultivate an atmosphere that embraces flexibility and communication. To do so, establish a process to receive routine feedback from

staff members about the workflow. This can be facilitated by encouraging staff members to pencil in changes on their desk procedures, pending formal document revisions. Flexibility and communication can also be facilitated by setting aside regularly established times to review workflows (e.g., the close of the fiscal year, calendar year, staff appraisal period).

The time period during which the process described in previous text is carried out will vary based on a number of factors. These factors include the timeline within which the new process must be ready and the time that can be devoted to the workflow analysis.

Analyzing E-Resource Workflows

Although the complexity and dynamic nature of e-resources pose great challenges to workflow managers, these resources are maturing and a library's accumulating expertise provides a solid foundation upon which more productive and responsive workflows can be built. Carrying out such workflow enhancements requires the manager to collaborate with a wide range of personnel. Indeed, from selection to access, e-resources have ushered in an evolution that requires partnership among many units within a library. Acquisitions personnel now routinely collaborate with librarians and staff members in administration, reference, public services, and information technology regarding such e-resource-related issues as: budgets, business terms, trial access, pending license agreements, access parameters, and renewal terms. Moreover, multiple departments have a hand in the e-resource acquisitions process and must work in concert to interact with content sources, create access, and serve users. As Bosch et al. state, "traditional boundaries between technical services, collection development and management, information technology, and public services become blurred and indistinct when dealing with electronic information acquisitions."[4]

A thorough analysis of e-resource workflows requires many of the same tools and activities discussed earlier in the print workflow analysis, but there are also some important differences between the analysis processes. The personnel and tools include:

1. unit staff members plus staff members from the other units that play a role in the processing of e-resources;

2. documentation of existing procedures plus those for the print serial workflow;
3. sample titles and issues;
4. sample records in the technology used to manage the process;
5. a flow chart (such as the one in Figure 14.3).

This flow chart can be used to frame the entire process and highlight steps within this process. The following steps augment the print plan described in previous text in order to provide a method for managers responsible for e-resource processes to analyze existing e-resource workflows for inefficiencies. The outcomes also mirror those stated in previous text: to educate staff members and foster collaboration and communication.

1. Carry out step (1) from the previous section, "Analyzing Print Serial Workflows"
2. Given the span of e-resource processing, focus initial efforts. For example, target acquisitions of e-resources and schedule subsequent analyses to focus on renewals, broken URLs and troubleshooting, cancellations, and archiving
3. Analyze the demands and needs of the e-resource type at hand
   a. Review the workflow for any corresponding print processes and determine if they are useful for this exercise. It may be possible to adapt an established workflow for this purpose. Can the workflow be adapted to address the aspects of the e-resource not found in the physical material?
   b. What needs are consistent across formats? When it is not possible to adapt the print workflow, how will the audit trail function and how will statistics/metrics be collected?
   c. What electronic resource management (ERM) tools are available? What parts of the process will each tool address? What synergy can be realized outside of the tools? Can the use of reports from the Integrated Library System (ILS), ERM, link resolver, journal list, and in-house spreadsheets or databases complement the work performed? Who will use which tools and when? Should staff receive additional development, training, or access permissions?
4. Carry out steps (3) through (9) from the preceding section on print serials. Talk with the staff members of the serials department

and address concerns and align perceptions prior to entering into discussions with other units within the library. Doing so will establish consistency and eliminate confusion around the acquisitions process prior to addressing broader issues that impact the workflows of others.

This process will take longer for e-resources than for print serials because the revision of e-resource workflows requires collaboration among more personnel in order to carry out a more complex process.

Analyzing serial workflows to accommodate e-resource management may sound difficult to staff members. The manager can overcome these difficulties by organizing the analysis into manageable segments and addressing staff members' concerns related to the complexity of the plan, the number of steps in the workflows, the tools involved, and the analysis process.

EFFICIENCY FOR SERIALS

The next step on the path to efficiency is to plan ways to optimize the revised workflows. This step requires the manager to examine what structure can provide a framework upon which the acquisitions unit can both survive mounting virtual titles and respond to the dynamic nature of the e-resources format. The following two sections of this chapter suggest how staff development and process integration can be harnessed to maximize the efficiency of the resources dedicated to serial workloads.

Maximizing Staff

The first consideration in achieving efficiency throughout serial workflows is staff assignments. Small libraries often have one staff member dedicated to several processes. In contrast, larger libraries may have highly specialized staff members whose expertise is focused on carrying out just one task. Regardless of size differences in libraries, there is one similarity: occasionally the volume of items processed can prevent staff members from meeting the demands of their work. One way to resolve this problem is to analyze staff assignments and devise plans to address shortages and increase efficiency.

One effective means to create efficiency via staff is to train at least two staff members to perform each critical process in a workflow. Even when staff members are burdened with backlogs in high-volume areas, it is possible to cross-train for targeted tasks over time. When necessary, consider working across units and departments to cross-train staff with aptitudes and interests in serials work. Utilize the tools needed to train new staff: accurate documentation, planning, and time. The staff designated to serve in a back-up capacity for a critical process can be used to address routine backlogs that can arise from surges in workloads.

Serials units shaped solely around the electronic format may eventually be overwhelmed by the mounting work as more titles move to an electronic-only format and the print materials received by the library decrease in volume. When managers observe such trends in purchasing and collection composition, they should plan ahead in order to reallocate staff resources. One way to carry this out is to create a plan that aligns staff members with the work. Libraries with ample staff may need to redistribute labor to more accurately represent their changing collections.

Staffing issues that may serve as barriers to achieving efficiency include: staff grade, staff expertise, and existing workloads. Critical thought regarding library resources and creative innovation can be employed to devise ideas for how existing staff can be reassembled to perform new assignments. A suggestion for addressing staff grade and class distinction is to allow higher-ranked staff members to bear more responsibility for tasks like communicating, overseeing processes, and resolving problems. Routine productivity should not be allowed to stall due to personnel grade. When staff expertise is at issue, a suggestion is to create a staff cross-training plan to be implemented over a period of time that allows staff the comfort of learning new skills while performing existing work. In addition, managers may consider examining existing work for tasks that can be automated, reassigned to student employees, reassigned to temporary staff members, or eliminated altogether. When the serials unit has materials remaining from the last software implementation or ILS migration, then a manager could temporarily rally resources from other areas until the backlog is eliminated.

Library budgets determine the level of development that staff members receive. In the absence of funding that can meet the continuing education needs of staff, time and information become commodities. Staff members need access to information to stay abreast of the issues impacting the library and their work. Staff development is a sound investment of time and, when available, funding because it can empower staff members to develop expertise and an awareness of local exceptions. In the absence of formal development, departments and units should establish informal internal development mechanisms to educate, stimulate thought, and spark creativity in staff. The objective is to give staff an understanding of the vision of the library and enable them to operate from a point of strength in the face of persistent changes in the field, publishing industry, and technology.

Integrating Processes

Another suggestion for possible serials efficiency is serials process integration. This has limited application for specific serials department configurations and is mentioned here to stimulate other ideas. At the infancy of the electronic format, the infrequent electronic subscription was an exception that caused workflows to stall because it fell outside of the traditional print process. The composition of serials now run the gamut: in some libraries they are evenly encountered along with print while other libraries focus on print or e-resources. Where print and electronic formats are now more evenly distributed, the integration of their respective workflows could create new efficiencies. Both may share components of a workflow similar enough that staff with sufficient development, training, and education can perform either process. Depending upon the staff aptitude and the ERM tools at hand, some staffs can handle the transition from print to electronic. Serials staff, as process experts, can also have the institutional knowledge of the serials collection. Good, solid performance records by staff justify the effort needed to train them to handle serials titles, regardless of format. Implementing such a plan in a timely manner could make both the staff and the workflows responsive across time and technology. Should a library consider integrating its print and electronic processes? Or should it maintain units dedicated to a specific format? Effective analysis of staffing resources and workflow processes within the library can reveal the solution. If a library chooses an integrated approach, here are some suggestions that may facilitate the plan:

1. Identify all of the processes being integrated, who (the process expert) currently performs each, who will be trained (the cohort) to perform each, and in what order. The order of the integration should be determined by the staff members' skills, internal resources, and workload.

2. Identify the tools used and what additional development and training may be needed. Will staff learn other ILS modules? Do they have rights to the ERM? Will they work with other departments to use the link resolver or journal list? Do they have access to status reports that will complement the tools?

3. Consider the current volume of work and how much time may be needed to perform the integration. Depending upon the time of year, focus integration efforts on ordering workflows or receiving/access workflows. Staff with large volumes of work to perform will need a slow, steady transition between old and new workflows at a rate that does not become counter-productive. Communication is important; install a feedback mechanism (e.g., a periodic meeting, so that managers and staff remain synchronized).

4. Conduct routine meetings about this process so that open communication can be exchanged between managers, the process experts, and the cohorts receiving training. A separate meeting may be needed periodically to address training issues or morale. Managers will need to be engaged to talk with staff about the policies that guide the process and generate a team spirit around the unit/department's objectives.

5. For the sake of expediency, let the process expert colead the team and train the cohort(s) until the cohort feels more comfortable. The amount of time will vary between cohorts and may be impacted by the availability of the type of work needed. Managers may have to troubleshoot the processes that exceed the expertise of the process expert. This is the first phase of the training.

6. Phase two of training is to let the cohort work with the support of the process expert until comfortable. Consider training complete when: (1) each task in the process is learned, (2) the cohort can answer his or her own questions, and (3) when the process expert is no longer needed at the workstation.

Executing workflow integration is as subjective as library staff configurations and workflows themselves. This topic is one example of serials efficiencies and could be a point of discussion for any serials department reviewing workflow and looking for avenues to maximize the use of staff and aligning resources with the changing composition of the collections.

CONCLUSION

As libraries address new competition for the patronage of users, success in the future depends upon the ability of libraries to design responsive workflows, achieve new efficiency and speed, and remain competitive in a crowded information marketplace. To meet these goals, serials managers are striving for efficiency, stability, economy, and flexibility in their workflows. In the life of serials, e-resources challenge managers to respond to mounting workloads and the dynamic nature of the format. In the meantime, print workflows can tie staff to legacy processes that are unresponsive to the changing composition of the collections. Despite the subjective nature of serial workflows for an individual library's set of circumstances (e.g., staff size, budget, collection size, and technology) every serials department can benefit from understanding issues that impact workflow. Departments can also benefit from periodic serial workflow analysis and work toward efficiency throughout the process of acquiring serials and establishing access.

NOTES

1. Jon Pyke, "What's Happened to Workflow? It's now Business Process Management!" *Information Management & Technology* 35 (2002): 254.

2. Kimberly Parker, Nathan D. M. Robertson, Ivy Anderson, Adam Changler, Shaon E.Farb, Timothy Jewell, and Angela Rigio, "Appendix B: Workflow Diagram," *Electronic Resource Management: Report of the DLF ERM Initiative* (Washington, DC: Digital Library Federation, 2004), 1.

3. Paoshan W. Yue and Rick Anderson, "Capturing Electronic Journals Management in a Flowchart," *Serials Review* (2007): 6 (forthcoming).

4. Stephen Bosch, Patricia A. Promis, and Chris Sugnet, *Guide to Licensing and Acquiring Electronic Information* (Lanham, MD: Scarecrow Press, Inc., 2005), 33.

Chapter 15

Issues in E-Resource Licensing

Jill E. Grogg
Selden Durgom Lamoureux

INTRODUCTION

With a growing percentage of collections budgets spent on e-resources, licensing has become a cornerstone of e-resource acquisition and management. Whereas libraries once purchased discrete containers of information in physical form, an explosion of networked e-resources—such as e-journals, aggregated article databases, and e-books—introduced a new element in the acquisitions process: licensing.

Licenses redefined the relationship between the publisher and the library and supplanted the role of copyright law. In the print environment, copyright law established the rights and expectations of the end user, the library, and the copyright holder. Not all publishers felt the need to create a license to cover materials published electronically, but many publishers considered copyright law insufficient in the digital environment, and turned to contract law to better protect their investment and more clearly define expectations of both parties. A key difference between contract and copyright, of course, is that contract terms can vary infinitely, and this fact has created special difficulties for both publishers and libraries. Standardization of license terms, an early prediction at many library and publisher conferences, has never materialized.

However, librarians today are seeing enormous and fundamental changes to licensing as the information and publishing industry moves to electronic delivery of scholarly content. It is possible for libraries to influence some of these changes, while other issues will require technological solutions not yet available. To the extent that libraries can use the license to influence the direction these changes take, they should.

While this chapter's focus is not on what to negotiate in a license, a few key issues that licenses commonly address are worth mentioning. Of special interest are terms in a license that have the potential to alter the permissions copyright law provided for print. A second set of issues may not have been previously addressed by copyright law but, nonetheless, has the power to change the traditional role of the library as we move from print to electronic. Three issues stand out: Interlibrary Loan (ILL), perpetual access, and archival rights. A library must first decide how important these issues are to its mission, and then ensure it licenses to maintain the rights librarians value.

ILL. In a license, it is possible to redefine or eliminate the ILL provisions of copyright law. While not all publishers try to do so, the library that is considering online-only should determine how important ILL is to it and license the rights needed. Some language to be on the alert for: prohibiting any ILL; limiting ILL to the country in which the library resides; and withholding permission to use secure transfer technology (e.g., Ariel). This is one issue that libraries have the ability to influence through licensing.

Perpetual access to subscribed content: This is a more difficult issue to resolve, as it has potential open-ended costs for publishers. In the print environment, a journal was the library's until deliberately discarded, lost, damaged, or stolen. The publisher's obligation to maintain the print issue ended with distribution. That is not the case in the digital environment. Once a library has cancelled a subscription, publishers must act to either: maintain a site (and manage subscriber rights to it), provide delivery of the content (e.g., CD-ROM, tape, e-mail), or allow the library to harvest, store, and display the content to patrons. Libraries that have a commitment to provide scholarly content to future scholars should be careful to include perpetual access rights in the license. While there are unknown future costs and technologies that will influence meaningful perpetual access, if it is important to the library's mission, the library should preserve that right in the license.

Archival access: This is a much more difficult issue to solve, as libraries and publishers are still in the early stages of developing viable solutions to the problem of how digital scholarly content will be preserved and who will be responsible. There are, however, several initiatives that should hopefully provide a good solution or from which a good solution will emerge: Lots of Copies Keep Stuff Safe (LOCKSS), Close LOCKSS (CLOCKSS), Portico, and trusted third parties such as the National Library of the Netherlands. Cost and technologies are two very large unknowns for libraries and publishers alike, but it is still advisable to include language in a license that indicates a library's interest in a reliable and affordable archive.

The challenge for libraries is to determine which, if any, of their core missions are affected by the three aforementioned issues and move to either assert the right to perform those missions, or reenvision their role. The intellectual process needed to make these decisions hinges not only on librarians' knowledge of the university and library mission but also on their understanding of both former and current initiatives in licensing.

In the past ten years, libraries and publishers have accumulated a considerable amount of experience with licensing. From early workshops that helped educate librarians about contracts to current initiatives to create new management systems to track license terms, libraries, publishers, and subscription agents have all reacted to make sense of and manage licensing in an evolving publishing model. This chapter takes a brief look back, and a hopeful look forward, at licensing e-resources.

EARLY INITIATIVES

Initially, licenses were negotiated for several formats, such as remote Internet access, CD-ROM, or tape. However, the past several years have seen the overwhelming preference for remote Internet access and the increased rarity of CD-ROMs or tape, except as vehicles for perpetual access rights. While the number of formats may have declined, the number (and frequency) of occasions for licensing has flourished. Licenses may need to be reviewed yearly, particularly in the case of leased information, or librarians may only have to negotiate licenses once, at the initial point of purchase.

In addition, companies have begun to cater to libraries that have one-time monies to spend. Thus, librarians have experienced a proliferation of one-time purchases of electronic content, such as the one-time purchase of specialty digitized collections (e.g., the Eighteenth Century Collections Online, from Thomson Gale) or the one-time purchase of e-journal package backfiles—which often require separate license agreements in addition to the previously negotiated license for ongoing journal content. Regardless of the format of electronic information or the nature of the product, the fact remained that licensing had become an integral task for librarians. David C. Fowler summarizes the evolution of licensing in libraries, noting, "the history of licensing for e-resources utilized by academic libraries, though short, has been a complex and, many would argue, a long and tortuous process."[1]

Publishers created licenses, and also offered the first "model" licenses, but many librarians and consortia were at the forefront in the early to mid-1990s when e-resource licensing emerged as a significant issue for libraries. These early innovators paved the way for the countless others who were struggling with how best to navigate the new licensing waters. A prime example is the LIBLICENSE project, spearheaded by Ann Okerson at Yale University. According to Okerson, "in 1996, CLIR [the Council on Library and Information Resources in Washington, D.C.] made the first of two grants to Yale University to create and launch the educational Web site . . ."[2] Since 1996, LIBLICENSE and its accompanying listserv, LIBLICENSE-L— available at http://www.library.yale.edu/~llicense/index.shtml—have grown into an invaluable source of licensing information. Available on the LIBLICENSE site are model licenses, the LIBLICENSE software, licensing terms and descriptions, a licensing vocabulary, a bibliography of useful articles as well as links to licensing resources, publishers' licenses, and links to national site and developing nations license initiatives.

Another leader in the licensing arena was and continues to be Trisha Davis of Ohio State University. Davis has written a number of articles and book chapters about licensing, and with others, developed a series of continuing education workshops about licensing for the Association of Research Libraries (ARL)/Office of Leadership and Management Services (OLMS). The ARL/OLMS licensing workshop served a critical purpose for librarians: to train them how to understand and negotiate licenses.

Many librarians do not receive such training in library school, so initiatives such as LIBLICENSE and the ARL/OLMS workshops emerged to fill this gap. As of this writing, the ARL/OLMS original workshop is no longer being offered, but LIBLICENSE is alive and well and other continuing education opportunities for librarians looking to enhance licensing skills are abundant. Many others, such as Ellen Finnie Duranceau at the Massachusetts Institute of Technology, and John Cox Associates, Ltd., were also key players in developing licensing education, commentary, and initiatives that are still important today.

MODEL LICENSES

Though the impetus for licensing came from publishers, this was not an inexpensive undertaking on their part, nor was it initially clear what to include in a license. In an effort to bring some standardization and structure to licensing, John Cox Associates pioneered a "model" license that offered publishers a set of terms and several variations of language to choose from within each term. This structure allowed publishers to build a sound license and offered the industry some standardization while avoiding any real or apparent collusion on the part of competing publishers. Though the Cox model license was not universally adopted, it still serves as the basis for many licenses offered by publishers today.

The challenge, of course, was to satisfy publishers' concerns about risk without overriding the needs of libraries and their parent institutions. As a result, several other model licenses were also developed early on. The United Kingdom's Publishers Association (PA)/Joint Information Systems Committee of the Higher Education Funding Councils (JISC) created an early model license,[3] and LIBLICENSE, in partnership with several other organizations, developed two other model licenses.

According to the LIBLICENSE Web site,

> the Liblicense Standard Licensing Agreement is an attempt to reach consensus on the basic terms of contracts to license digital

information between university libraries and academic publishers. Sponsored by the Council on Library and Information Resources, the Digital Library Federation and Yale University Library, it represents the contributions of numerous college and university librarians, lawyers and other university officials responsible for licensing, as well as significant input from representatives of the academic publishing community.[4]

LIBLICENSE points to a model "short" form license, which is only one page in length and attempts to simplify the licensing process for both publishers and librarians.[5] LIBLICENSE also points to an institutional model license,[6] which is lengthier but still offers librarians and consortia a place from which they can begin. They can use the model license to review and compare a publisher-provided license or develop a standard institutional or consortium license of their own.

STANDARD LICENSES

The term "standard license" is used here to refer to a license developed by librarians, either at a specific institution or as part of a consortium or other group. Duranceau, in her Winter 2003 Electronic Journal Forum column in *Serials Review,* describes a pilot study at Massachusetts Institute of Technology (MIT) Libraries in which the librarians developed a standard license based on the Yale model license. They used this MIT Standard License for individual e-journal purchases. Duranceau explains the impetus for the project: "during the winter of 2001 and spring of 2002, the MIT Libraries realized that the pressure of negotiating license agreements for so many individual e-journals had reached a critical point and that creative solutions to the licensing bottleneck were required to prevent long delays in establishing e-journal access."[7] Duranceau goes on to detail the success rate of the official pilot: 35 percent accepted the MIT license outright; 24 percent accepted the MIT license with some revisions; and 41 percent rejected the MIT license.[8] All in all, Duranceau reported that using the standard license worked well and that she was encouraged by her success and the experiments at other libraries, such as Yale University.[9]

Though not widely used, standard licenses have grown in popularity, and a number of other institutions as well as state, regional, and national consortia have developed their own versions. For example, Harvard University Library has made available its "Licensing Electronic Resources at Harvard University: Guidelines for Vendors," which contains detailed explanations of Harvard's required licensing provisions.[10] Stephen Bosch, in a 2005 article in the *Journal of Library Administration,* gives an overview of the historical and current use of model or standard licenses. In terms of national model licenses, he mentions the United Kingdom's National Electronic Site License Initiative (NESLI), available at http://www.nesli.ac.uk and Canada's *Consortia Canada* model license. Bosch also points to regional and state-wide licenses, such as the Northeast Regional Libraries and the California Digital Library.[11]

Bosch emphasizes that "model licenses were not created as part of a fad," but were created in "direct response to the need to bring the licensing process under control."[12] Indeed, many a library has an incredible backlog of licenses, and in the early days of the license explosion, many libraries did not have a coordinated licensing workflow or a single person dedicated to managing the licensing process. Therefore, the model or standard license became a way for a given library to rein in its licensing process and create standardized workflows.

Furthermore, Bosch offers the following advice for those using model or standard licenses: "some organizations may be large enough that they really can offer their model [or in the terms defined here, standard] license as the starting point in negotiations, but that scenario is probably the exception, not the rule. This does not diminish the value of a model document. Even if a supplier does not work from the document, the model will make the work go more easily."[13] Moreover, even if neither the library nor the vendor actually uses the model document, this still does not diminish the value of the document itself. The very process of creating or adapting a model or standard license for a particular institution's or consortium's needs forces the organization to "develop understanding of licensing in the organization."[14]

Finally, even for the smallest or most understaffed organization, reviewing a preexisting model or standard license can be a starting point for creating a basic institution-specific checklist—a checklist detailing what is and is not acceptable as well as what is desirable. The use of model and standard licenses has a variety of direct and indirect benefits and drawbacks, but ultimately, their continued use will only force the library and the supplier to come to greater understanding of common terms and common needs.

OTHER EXPERIMENTS IN LICENSING

Publishers and libraries are not the only interested parties active in licensing standardization efforts. Subscription agents have also been active in offering solutions. Bosch notes that in 2000, five subscription agents arranged for John Cox Associates and librarians to work together to create a series of model licenses, "one each for academic libraries, academic consortia, a public library license, and another one for both corporate and special libraries."[15] These model licenses have been revised and are now freely available in version 2.0 at http://www.licensingmodels.com/.

Efforts to try to standardize licenses were born from a very real need on the part of both publishers and librarians to streamline the licensing process. Whether starting with a publisher-supplied license or a library standard license, the hope was to reduce the amount of time spent negotiating and processing the license. Nevertheless, no single solution has emerged, and librarians, publishers, and subscription agents continue to develop new tools to manage the process, and to search for creative alternatives to current practice.

ALTERNATIVE TO LICENSING

An alternative to the labor-intensive licensing process is to do away with licensing altogether. Though there are many instances where the business model may be so complex that both parties will still feel the need for a contract, in many other cases the terms for e-resources are straightforward, and the expectations of both parties have come to be broadly understood and accepted. What has been lacking is an articulation of those shared understandings.

In 2005, the idea for a "Best Practice" document, described in Informed Strategies "Streamlining the Supply Chain,"[16] was presented at several library conferences, and in the fall of 2006, a small group with representatives from libraries, publishers, and subscription agents met to draft a document that would describe these shared understandings.

What has emerged from that meeting is a National Information Standards Organization (NISO) Working Group (working title, SERU: Standard E-resource Understanding) that is in the early stages of creating a document that could serve as the standard for transactions between libraries and publishers. Currently, this initiative is in its infancy. It remains to be seen whether it will develop and take hold. At stake is the elimination of negotiated contracts from our interactions with publishers and all that it implies for both the publisher and the library.

LICENSE EXPRESSION WORKING GROUP

All interested parties can be heartened by an ever-increasing number of content providers and librarian collaborations. An example of this is the License Expression Working Group (LEWG), which formed in 2005 as a joint effort of the National Information Standards Organization, Digital Library Federation (DLF), EDItEUR, and the Publishers Licensing Society (PLS). An outgrowth of the extremely successful Electronic Resources Management Initiative (ERMI), LEWG is charged with developing a single standard for the exchange of license information between publishers and libraries. More specifically,

1. Monitor and make recommendations regarding the further development of standards relating to e-resources and license expression, including but not limited to the ERMI and EDItEUR work;
2. Actively engage in the development of the ONIX license messaging specification.[17]

While the group has a fairly specific charge of creating standards for the exchange of license information among particular systems, such as electronic resources management systems, other benefits have emerged as well.

Many of the licensing initiatives have experienced unforeseen benefits. For example, as described earlier, whether an institution or organization actually uses a model or standard license does not diminish the value of reviewing such a document, as it forces the organization to review its own licensing needs. Similarly, if LEWG can come to a consensus about the common expression of elements in a license, then perhaps we can move closer to a less-intensive negotiation process on the part of both the library and the publisher. At the very least, if the concepts and vocabulary are standardized, then that facilitates the automatic population of ERM systems, which represent the best kind of knowledge management for librarians and end users alike.

Several members of LEWG—Tim Jewell (University of Washington Libraries), Trisha L. Davis (Ohio State University Libraries), and Diane Grover (University of Washington Libraries)—have conducted workshops about mapping license language for ERMs at library conferences such as the North American Serials Interest Group Conference in 2006 and the American Library Association Annual Conference in 2006. A report by Jill Grogg detailing the content of the workshop is available in the 2006 North American Serials Interest Group (NASIG) proceedings, published in *The Serials Librarian.*

END-USER EDUCATION AND OTHER ISSUES

When discussing licensing issues, the focus is often on the publishers and the librarians, but in the final analysis, it is often the end user who most benefits or suffers from a poorly constructed or negotiated license. It is also the end user who is most likely to commit infringement of terms of use, so his or her education about the terms to which the institution has agreed is paramount. Jill Emery summarizes the past and current practices for educating users about licensing terms in 2005: "we need to tell end-users more than just who can access what—we need to explain specific limitations that may be associated with each resource."[18] Emery goes on to explain that current tools for conveying licensing terms, ERMs and Online Public Access Catalogs (OPACs), should not be viewed as the complete answer.

Emery encourages electronic resources librarians "to provide more in-house training to their public services staff to ensure the terms of use are being taught in instruction sessions and pointed out during extended reference interviews."[19] In essence, librarians must be proactive in translating complex licensing terms both in-house and to their respective constituencies. This is not only true from an ethical standpoint, but also from a contractual standpoint as well. For example, many licenses contain language that requires the institution to take reasonable efforts to educate end-users about acceptable and allowable behavior.

Other licensing initiatives and issues on which to keep a careful eye include usage statistics (Project COUNTER, http://www.project counter.org/) and digital preservation and electronic information archival terms (LOCKSS, http://www.lockss.org/lockss/Home; CLOCKSS, http://www.lockss.org/clockss/Home; and Portico, http://www .portico.org/). In addition, the open access movement will continue to inform discussions of current publishing practices and pricing models, and groups such as Scholarly Publishing and Academic Resources Coalition (SPARC, http://www.arl.org/sparc/) will remain important advocates for a functional scholarly communication system.

CONCLUSION

As license review and negotiation were not traditional activities associated with librarianship and, therefore, traditionally not offered as courses in library schools, librarians faced with these tasks have been forced to educate themselves. Librarians quickly learned that basic knowledge of copyright law was a must, and workshops such as Laura N. Gasaway's "Copyright Law in the Digital Age" were critical. In addition, librarians also quickly learned that an accurate assessment of one's local environment was a necessary first step. While librarians are generous and share information about licensing practices and model licenses, each institution or organization will necessarily have its own unique attributes.

Librarians also discovered the importance of a good relationship with local legal counsel and the importance of enhanced negotiation skills. Continuing education opportunities remain critical for those who negotiate and review licenses, and these opportunities include a preconference sponsored by the American Library Association's Association for Library Collections and Technical Services (ALCTS) and led by Trisha Davis and Becky Albitz of Penn State University Libraries and the Electronic Resources and Libraries Conference held in Atlanta, Georgia, for the first time in 2006. Ultimately, librarians ascertained that any negotiation for the purchase of an e-resource was about much more than money; it was about providing the best possible access for his or her particular constituency.

Perhaps Fowler, in his historical review of licensing, says it best: "a decade of experience with electronic licenses has created a set of shared experiences between libraries and publishers . . . most publishers now realize that librarians are better to have as partners than adversaries, and have worked to create much more palatable electronic licenses with which all parties can be happy."[20] Indeed, like many of the recent issues that have emerged from the explosion of e-resources—linking, cataloging, preservation, budgeting, selection, and much more—licensing has come a long way in a very short amount of time.

NOTES

1. David C. Fowler, "Licensing: An Historical Perspective," *Journal of Library Administration* 42, no. 3/4 (2005): 178.

2. Ann Okerson, "The LIBLICENSE Project and How it Grows," *D-Lib Magazine* 5, no. 9 (September 1999): http://www.dlib.org/dlib/september99/okerson/09okerson.html (accessed January 7, 2007).

3. Stephen Bosch, "Using Model Licenses," *Journal of Library Administration* 42, no. 3/4 (2005): 67.

4. LIBLICENSE, "CLIR/DLF Model License," (1996-2006), http://www.library.yale.edu/~llicense/modlic.shtml (accessed January 7, 2007).

5. LIBLICENSE, "Short Form Agreement, Version 1.0," (August 9, 2001), http://www.library.yale.edu/~llicense/shortform.html (accessed January 7, 2007).

6. LIBLICENSE, "Standard License Agreement, Version 2.0," (July 4, 2001), http://www.library.yale.edu/~llicense/standlicagree.html (accessed January 7, 2007).

7. Ellen Finnie Duranceau, "Electronic Journal Forum: Using a Standard License for Individual Electronic Journal Purchases: Results of a Pilot Study in the MIT Libraries," *Serials Review* 29, no. 4 (2003): 302.

8. Duranceau, "Electronic Journal Forum," 303.

9. Ibid., 304.

10. Harvard University Library, "Licensing Electronic Resources at Harvard University: Guidelines for Vendors," (January 2000), http://hul.harvard.edu/ldi/resources/vendor_guidelines.pdf (accessed January 7, 2007).

11. Bosch, "Using Model Licenses," 69.

12. Ibid., 70.

13. Ibid., 76.

14. Ibid.

15. Ibid., 68.

16. Judy Luther, "Streamlining the Supply Chain," Informed Strategies (March 2006), http://smartech.gatech.edu/dspace/bitstream/1853/10071/2/20060320_ BestPracticeOption.8.pdf (accessed January 7, 2007).

17. National Information Standards Organization, "License Expression Working Group," (2005), http://www.niso.org/committees/License_Expression/LicenseEx_ comm.html (accessed January 7, 2007).

18. Jill Emery, "Is Our Best Good Enough? Educating End-Users About Licensing Terms," *Journal of Library Administration* 42, no. 3/4 (2005): 38.

19. Ibid.

20. Fowler, "Historical Perspective," 196.

Chapter 16

The Activation and Maintenance of E-Journal Access

Patrick L. Carr

INTRODUCTION

After a library has acquired the right to access an e-journal, it must activate and maintain this access. Meeting these objectives is a challenging process. It requires a library's e-resource administrator(s) to develop detailed yet flexible workflows and to rely upon a wide and constantly changing array of tools and partners, many of which were nonexistent in a print-dominated environment. Kittie Henderson vividly describes the added degree of difficulty inherent in the management of e-journals when she comments that, "if print serials are like greased pigs, then e-journals are like greased pigs on speed."[1] Through a discussion of the activation and maintenance of e-journal access, this chapter describes how a library can grab hold of these speed-addled swine, wrestle them to the ground, and lock them up in their pen. Put more literally, it provides an overview of the tasks, tools, and partners involved in the administrator's efforts to make sure that e-journal access is successfully provided to users and managed by the users' library.

THE ACTIVATION OF E-JOURNAL ACCESS

The term "activation" is used frequently in discussions of e-resource management, but the meaning of this term varies according to the

Managing the Transition from Print to Electronic Journals and Resources

context in which it is placed. In this chapter, "activation" is defined broadly. It constitutes all actions that a library must take to make an e-journal accessible to its users once the library has acquired the right to access this resource—either through a direct subscription, a consortial partnership, or some other means. Given the broadness of this definition and the variety of tools that publishers and libraries have in place to make e-journals accessible, a distinguishing element of activation workflows is clearly their diversity. Despite this multifarious nature, at least one fundamental distinction can be drawn when describing workflows for the activation of any e-journal at any library. This distinction consists in the procedures necessary to make sure that

1. a library's users have access to an e-journal on the resource's external access platform;
2. an e-journal's accessibility is reflected in the appropriate internal information retrieval systems that the library provides for its users.

To reflect this distinction, discussion of activation workflows is divided into two sections. The first section focuses on what this chapter will term external activation: procedures necessary to establish that e-journal access exists. The second section focuses on internal activation: procedures necessary to ensure that a library's information retrieval systems reflect an e-journal's external access.

External Activation

At a time when print was the dominant format in which libraries acquired journals, the claim sufficed as a straightforward action for communicating to a publisher that subscribed content has not been received. With e-journals, however, the process of communicating that a library does not have access to subscribed content has become at once more urgent and more complex. The process has increased in urgency because—unlike most claims for print journals, which are for particular issues of the journal's purchased content—a "claim" for an e-journal often aims to establish access to all of the content subscribed to by the library; in other words, with e-journals, a library frequently has access to either all subscribed content or none of it. This added degree of urgency is accompanied by an added degree of complexity because of the diverse assortment of platforms that publishers use to provide access to e-journals. While most claims for print journals can be handled from start to finish by a library's subscription agent, the services that agents can provide in the external activation process are more limited. The extent of what the library must do in this process largely depends upon the nature and polices of the e-journal's access platform.

Sometimes online access is provided through a platform created by the publisher itself. In cases in which the publisher is large (e.g., Blackwell, Elsevier, Springer, Wiley, etc.) this arrangement is usually advantageous: the publisher can determine directly from its records what e-journals a library should be given access to and then effectively manage this access with little assistance. Perhaps the greatest problem with large publishers is that, when a library does identify and report an access problem, the publisher—either because it is too busy or because of a lack of organizational communication and coordination—is not always able to address and resolve the problem in a prompt manner. Indeed, in some cases, administrators are forced to repeatedly send follow-up communications to the publisher to have an access problem resolved. Fortunately, some of the problems that a library encounters with e-journal access on a large publisher's platform are such that it can draw upon the experiences of other libraries in order to work toward a resolution. For example, on May 30, 2006 there was a posting on the Liblicense listserv (http://www.library.yale.edu/~llicense/index.shtml) in which a library indicated that it had received a rash of security alerts from Blackwell that warned of excessive downloading by Web crawling robots (i.e., spiders) on the publisher's platform, Blackwell Synergy, and indicated that, as a result, one of the library's IP ranges had been shut off from access. An internal investigation by the library, however, did not support the claims made in the security alerts. This posting received numerous responses from other libraries experiencing similar problems with the publisher. Through these communications, the libraries were able to share ideas about local solutions and place pressure on the publisher to develop a global solution.[2]

When the publisher is small in size, an arrangement in which online access is provided through the publisher's own platform can be more problematic. There are several reasons for this. First, it means that the e-resource administrator must take the time to learn the idio-syncrasies of the publisher's activation procedures and modules. Lo-cating the e-journal's access point on the Internet, determining how to carry out online activation, and then getting a representative from the publisher to respond to the administrator's claim for online access are among the challenges that must be overcome. Although there are some small publishers that will allow a subscription agent to activate e-journal access on a library's behalf, many require the library itself to activate its access. Indeed, it is often the case that a library's sub-scription agent can only facilitate this process by providing infor-mation about how to activate access and giving the codes and ID numbers that the publisher requires in its activation process.

Further complicating the activation of e-journals on a small pub-lisher's platform is the requirement that some of these publishers have for a library to accept the terms of a license agreement. Although li-cense agreements are typically a component of the acquisitions pro-cess rather than the activation process, in cases in which online access comes "free" with a print subscription, a license agreement—fraught with all of the issues discussed in the previous chapter—is sometimes required. Since the library has already acquired the e-journal, the e-journal's publisher usually has little to gain by agreeing to work with a library on the terms of such a license agreement. Accordingly, some libraries have reported that the problematic conditions of these licenses, combined with the publishers' unwillingness to modify them, can result in the decision not to activate access to an e-journal even though it is included in the library's subscription.[3]

An additional reason why activating online access through an inter-face created by a small publisher can be problematic is that the proce-dures and modules that have been developed for this purpose are not always geared toward the needs of libraries. Indeed, figuratively— and sometimes literally—the publisher and the library's e-resource administrator do not speak the same language. Many small publish-ers, for example, have developed an e-journal authentication process with individuals or corporate entities rather than institutional subscrib-ers in mind. One telltale sign that such an orientation exists is that the publisher does not allow IP authentication but instead requires that users gain access to e-journal content by entering a username and pass-word. While this arrangement may be acceptable to the individual or corporate subscriber, the requirement of gaining access via a password is the source of much ire for the e-resource administrator. In many cases, the administrator may determine that it is simply unfeasible for a library to provide this form of access to an e-journal. Indeed, for some of these e-journals, the user who signs in with a username and password is given all of the privileges of the manager of the subscriber's account. This means that the user may be able to modify the library's contact information, access codes, and mailing address, and even act on the library's behalf in order to begin new subscriptions to the publisher's products. Even if the user is not given the privileges of an account manager, username/password access remains highly problematic. Difficulties related to this means of authentication include locating passwords that must be retrieved from an insert in the print version of the journal and keeping track of an array of passwords that constantly change and/or are an incomprehensible jumble of numbers and letters.[4]

Many publishers partially alleviate libraries' activation woes by partnering with a third-party hosting service, or multipublisher platform, to provide online access to e-journal content; examples of these services include Highwire, Ingenta, and MetaPress. From the perspective of a publisher, there are a number of compelling reasons to employ a hosting service. By capitalizing on the ready-made audience of users who are familiar with a hosting service's interface, a publisher can expand its Web presence and increase the usage of its e-journals' content; this increased usage can take the form of pay-per-view transactions, downloads of complimentary articles, and access by subscribers. In addition, such a service frees the publisher from the financial and technological burdens of developing and maintaining its own platform for providing online access to subscribers. Although a hosting service may mean less control over how a publisher delivers e-journal content, the benefits of a secure, user-friendly, and highly recognized platform for access are substantial.

For a library, a hosting service can also be advantageous. By providing a level of consistency, they can do much to streamline activation workflows. In some cases, a subscription agent can act on a library's behalf to activate access through a hosting service; EBSCO Subscription Services, for example, will activate online access to all of a library's subscriptions included on the Ingenta and MetaPress platforms

(MetaPress, it must be mentioned, is a division of EBSCO Industries, Inc.). When a subscription agent cannot perform this task for a library, activation typically entails that the e-resource administrator log in to the hosting service's administrative module, enter required fields of information (e.g., the library's costumer number, IP addresses, and administrator contact information), and then monitor to ensure that online access has been granted. However, third-party hosting services are not without their problems. Owing to their subscription agent-like position as a "middleman" between publishers and subscribers, occasionally these services do not recognize that a library holds a current subscription to one or more e-journals and, accordingly, deny the library's claim for activation. In such instances, it often falls on the e-resource administrator's shoulders to contact the hosting service, its subscription agent, and/or the publisher in order to resolve the problem.

Internal Activation

Earlier in this chapter—and, indeed, throughout this book as a whole—the reader will encounter statements insisting that, with the transition from print to e-resources, library procedures have become far more complex and difficult. However, when it comes to the internal activation of e-journals—namely, procedures designed to ensure that access to an externally activated e-journal is reflected in the information retrieval systems that are provided for a library's users—this often-repeated statement is, in some respects, incorrect. The reason for this is that the tasks necessary to do what in a print-dominated environment is equivalent to activating internal access—for example, taking the journal out of its mailing package, checking it in, assigning it a call number, and shelving it—are all focused on a specific print piece that is received by a specific library at a specific time. With e-journals, however, libraries no longer need to deal with these specifics. Indeed, despite the extraordinarily dynamic nature of an e-journal's content, the content that any one library subscribing to the e-journal has access to at any one time is, in general, uniform with the content that all other libraries subscribing to the e-journal have access to at that time. Moreover, an addition to an e-journal's externally accessible content usually does not necessitate a specific action on the part of a subscribing library in the way that, for instance, a newly received print issue does; indeed, with most e-journals, internal activation only requires action at the journal-level or—in the case of aggregated or publisher packages—at the package level.

The implication of these changes is that a library can dramatically simplify its internal activation procedures by providing users with e-journal access points that are powered by a single, externally maintained knowledgebase. With such a knowledgebase (offered by such companies as Ex Libris, Serials Solutions, TDNet, and most major subscription agents), a library does not, for example, need to worry about updating each of its access points if a new title is added to an aggregated package of e-journals to which it subscribes. In such a case, the externally employed administrators of the knowledgebase would act on behalf of this and many other libraries by updating the knowledgebase—and, in doing so, updating the library's access points—to reflect the access that the library has gained. By completing what usually amounts to just one simple action—namely, "switching on" an externally activated e-journal or e-journal package in the knowledgebase in order to indicate that the library holds a subscription—a library is able to internally activate the resource from numerous points within the library's Web presence and beyond. Of course, the most common examples of these points include A-to-Z title lists, metasearch applications, and—through an OpenURL link resolver—citations to the content of an e-journal that appear within many other e-resources that are accessible to the library's users.

A library's online catalog can also function as an important retrieval and access tool for e-journals. If a library subscribes to a MARC record service that is powered by its externally maintained knowledgebase, it can regularly add or modify records for acquired e-journals in its online catalog. Moreover, the libraries at some institutions, such as the University of Washington and Oregon State University, have reported implementing workflows in which they can customize and enhance e-journal access in the catalog by integrating local cataloging procedures into their MARC record services' regular updates.[5] Still other libraries do not use a MARC record service but instead have procedures in place through which they manually catalog certain categories of e-journals to which they have access.

In addition to making sure that an e-journal is included within a library's various access points, internal activation also entails insuring that the access provided to an e-journal includes all appropriate portions (and only those portions) of its user community. Indeed, just as a publisher must have a means to determine if a library should be able to activate external access to an e-journal, a library must make certain that the internal access it provides reaches all users who are stipulated in its license agreement but prohibits access to those users that the agreement excludes. The nature of the procedures that a library must carry out here depends upon how the library authenticates users. The authentication strategy that is utilized by most North American academic libraries is IP authentication for on-site users accompanied by proxy server authentication for off-site users. In most cases, this authentication strategy is a simple matter for a library and may only involve registering an e-journal's domain on the library's proxy server—an action which allows registered off-site users access to the e-journal.

In other cases, however, authenticating users may be more complicated. For example, an e-journal to which only certain portions of a library's user community are permitted access (e.g., only those users accessing the e-journal from a university's main campus or only those users accessing the e-journal from designated workstations) requires that the e-resource administrator—oftentimes collaborating with systems specialists in the library—take special steps to both comply with these restrictions and communicate their nature to users. An e-journal that does not allow IP filtering and instead requires the entry of a username/password presents another challenge to the administrator attempting to guarantee that internal access is provided to (and limited to) appropriate users. For some libraries, the imperfect solution to this problem is to integrate notes concerning username/password information within the access points that it provides for an e-journal. Other libraries, however, believe that this process is too labor-intensive and insecure; a 2004 book chapter, for example, states that the University of Californian Los Angeles, (UCLA) Biomedical Library requires users to request username/password information at its reference desk.[6]

Another important issue that an administrator must consider when activating an e-journal's internal access is which platform(s) for the e-journal the library's access points should link to. Indeed, the content acquired through one subscription to an e-journal may be externally accessible from platforms created by the publisher, the library's subscription agent (examples of these platforms—sometimes referred to as gateway services—include EBSCOhost Electronic Journal Service and SwetsWise), and a third-party hosting service. A library can, of course, opt to provide its users with access to an e-journal through all of its available platforms. Perhaps the primary advantage here is that it maximizes the users' options for accessing an e-journal, allowing them to decide which of the platforms' interfaces they prefer and, in some cases, giving them an alternative platform if they experience a problem accessing content from their initial choice. Among the problems with internally activating all possible platforms are that it can inflate the library's title count. This, in turn, results in a large and unwieldy list of e-journals that is more labor intensive for the library to effectively manage and baffling for the user to navigate through. The library choosing to provide internal access through just one of these platforms must compare the advantages and disadvantages of each option—a complex undertaking that will only be discussed briefly here. The platforms offered by hosting services and subscription agents have the advantage of increasing the consistency of the interfaces that a library provides for users to access e-journals. Furthermore, by using a subscription agent's platform, the library can integrate its access to e-journals with the information and modules that the subscription agent provides for the library to manage these e-journals. One disadvantage of the platforms offered by both hosting services and subscription agents is that, as middlemen between publishers and libraries, their information about the content that libraries should have access to may be inaccurate and/or out-of-date. Moreover, many of the e-journals on a subscription agent's platform may not even be accessible on this platform; instead, users may have to search on the platform's interface for a hyperlink that takes them to the publisher's platform—an arrangement in which the subscription agent's platform is rendered a confusing and unnecessary intermediary page through which the user must navigate to reach an e-journal's content.

MAINTENANCE OF E-JOURNAL ACCESS

After a library has activated its access to an e-journal, it must work to be sure that this access is maintained. With print journals, the maintenance of access is achieved in a library's Integrated Library Service (ILS) through a widely used three-step process:

1. The library creates predictions for issues that it should receive.
2. It uses these predictions to check-in issues that it does receive.
3. It periodically runs a claims report to generate a list of all issues that should have been received but were not.

In the rapidly changing landscape of e-journal management, no clearly defined process has not arisen that the majority of libraries rely upon to guarantee that access is maintained. Instead, libraries have developed a variety of procedures that are rooted in such unique factors as their human and technological resources, organizational structures, and users' needs. In general, however, the procedures that libraries rely upon to ensure that e-journal access is maintained fall between two extremes.

One of these extremes is to apply to e-journals the same basic workflows that a library uses to guarantee access to print journals. In her 2006 North American Serials Interest Group (NASIG) conference presentation "Old Is New Again: Using Established Workflows to Handle Electronic Resources," Amanda Yesilbas describes how Florida Atlantic University (FAU) Libraries has successfully modified print workflows to maintain access to e-journals. Yesilbas explains that, after discovering that users did not have access to a significant number of subscribed e-journals, the Libraries determined that they needed to develop a systematic method to maintain access to subscriptions. The method developed was e-journal check-in. Here, staff members use issue predictions modeled on the print publication cycles of subscribed e-journals (excluding titles in aggregated packages) in order to regularly visit each e-journal's access platform and test it to verify that users still have access. According to Yesilbas, e-journal check-in at FAU Libraries has proven to be a manageable and proactive method of making sure that the Libraries promptly identify licensing, payment, and technical problems impacting e-journal access; in addition, e-journal check-in facilitates access troubleshooting by providing the Libraries with a clear record of when access to each subscribed e-journal was last verified.[7]

A markedly less systematic and proactive method for maintaining e-journal access is called waiting for users to complain. To follow this method, the e-resource administrator simply relies upon complaints from frustrated users in order to identify instances in which subscribed e-journals are not accessible through the library. Despite the obvious disadvantages of this method for maintaining e-journal access, many administrators, beleaguered by all of their other e-resource management responsibilities, have come to rely upon it.

Fortunately, between the two extremes of manual e-journal check-in and user complaints, there are opportunities for a library to stake out a middle ground that combines the systematics of the former method with the pragmatics of the latter method. For example, Yesilbas reports that one discovery she made from managing e-journal check-in at FAU Libraries was that a significant majority of e-journal access problems occurred in the first few months of each calendar year—the period in which their subscription agent renewed subscriptions with publishers.[8] Accordingly, rather than checking access in correspondence with an e-journal's print publication cycle, a library might abbreviate this process by only checking access during those first few months of the year when it is most likely that there will be a problem. Likewise, a library may be able to identify certain "bad apple" e-journal platforms that are more prone to access problems and then focus systematic efforts on checking the e-journals accessed through just those platforms. Lastly, ILS vendors are beginning to develop products that can automate e-journal check-in. Innovative Interfaces, for example, has developed Millennium Serials E-Checkin, a product that utilizes XML technology in order to transmit information concerning newly accessed e-journal content directly from the publisher to the library's Millennium server, where the resource's check-in record is updated.[9]

An additional strategy that an e-resource administrator can employ in order to ensure the maintenance of e-journal access is developing clear channels of communication with access partners. For example, one common reason why a library loses access to e-journals is that these resources change publishers and, accordingly, change access platforms. To identify when such changes occur, administrators can monitor relevant listservs and publisher announcements and take advantage of e-journal access reports offered by their subscription agents. Another common source of e-journal access problems is errors and updates in the knowledgebase that a library uses to power its internal access points. To identify these problems, administrators can monitor customer support listservs and carefully review the information included in the knowledgebase's update reports. For libraries in consortial partnerships in which all libraries have access to e-journals subscribed to by any one library, administrators need to be sure to communicate any new or cancelled subscription to all partner libraries.

EMERGING TRENDS IN THE ACTIVATION AND MAINTENANCE OF E-JOURNAL ACCESS

The future of libraries' efforts to activate and maintain e-journal access will likely be shaped by two conflicting trends in the publication and delivery of content. One of these trends is the growing share of the marketplace being staked out by large publishers. As publishing consultant October Ivins indicates, the financial and technological resources necessary for a small publisher of journals to make the transition from print to electronic publication is overwhelming and can often result in this publisher selling out to a larger competitor.[10] Although there are reasons for information professionals to be alarmed by this trend, its implications regarding libraries' activation and maintenance of e-journal access are positive. Indeed, with the growing dominance of large publishers and their accompanying "big deal" packages, publishers' access platforms and libraries' accessible e-journal content are becoming homogenized—a shift that results in the streamlining of a library's procedures for the activation and maintenance of e-journal access. In other words, as the e-journals of small publishers are bought by larger competitors, the platforms that a library uses for access become less diverse, less idiosyncratic, and, accordingly, less time- and labor-intensive to work with. Moreover, as libraries are enticed away from subscriptions to individual e-journals and toward subscriptions to established publisher packages that are also subscribed to by other institutions, it becomes less difficult for a library to manage the knowledgebase it uses to power its e-journal access points. Instead of working in this knowledgebase in order to individually search for and "switch on" each e-journal in a subscribed package, the administrator can simply "switch on" the entire package.

In a future dominated by large publishers, the role and value of the subscription agent as a partner in e-journal access and maintenance is being called into question. With respect to journal access, the core service that an agent has traditionally provided consists of eliminating a library's need to communicate problems with a diverse array of publishers, each with its unique claiming policies and customer service representatives. With the increased homogenization of means for e-journal access and the streamlining of activation procedures, the value of an agent becomes less apparent. Although agents are attempting to radically change their services in order to maintain their position as a vital stakeholder in the realm of e-journal access and maintenance, they are at a disadvantage due to their position as an intermediary and the ambition of some large publishers to enhance their services by working directly with libraries; for example, Tony McSeán, director of library relations at Elsevier, has argued that an agent does not contribute any value to the e-journal subscription process.[11]

The other trend shaping the future of libraries' efforts to activate and maintain e-journal access is the Open Access (OA) movement through which scholarly publications are made freely available online. Although there is wide debate about the nature of the OA movement and great uncertainty about its future, it is clear that, whatever shape this movement takes, it will have significant implications concerning how libraries provide and maintain access to e-journal content. Perhaps the most far reaching implication of the OA movement is that it may result in the dissolution of the journal title—or more specifically, the journal issue—as the entity determining the articles to which a library's users have access. For example, many publishers (including the American Chemical Society, the American Physical Society, Blackwell, Elsevier, Oxford University Press, Springer, and Taylor & Francis) are allowing the authors of articles in certain journals to pay a one-time fee in order to make an article freely accessible online. The result is "hybrid journals" in which some articles are accessible to a library's users and others are not. Moreover, through e-print archives and institutional repositories, users have free access to growing treasuries of publications which may not even be components of particular journals. At present, libraries lack the tools that could readily determine whether a particular publication for which a user is searching is freely available online. In the future, however, it is conceivable that the knowledgebases used to power internal access points will be reconfigured so that the article (perhaps identified through a Digital Object Identifier) rather than the journal title will become the basic unit through which access is identified.

CONCLUSION

The activation and maintenance of e-journal access is a challenging, multifaceted process. It constitutes the crucial final step a library must take to enable users to gain access to resources for which it has often taken great care to negotiate acceptable licensing and business terms. This chapter has presented only an overview of some of the most important tasks, tools, and partners that are involved in this process. As such, the effective activation and maintenance of e-journal access is clearly an achievement which is key to a library's successful transition from print to e-resources. Delving much deeper into this topic, however, would require the stipulation of a specific infrastructure for a library's management of e-resources. Accordingly, these more detailed processes are better addressed through case studies on the activation and maintenance of e-journal access at specific libraries—contributions which have appeared too rarely in the professional literature.

NOTES

1. Markel Tumlin, ed., "Is Check-In Checking Out?" *Serials Review* 29, no. 3 (2003): 225.

2. The full e-mail thread (subject line: "Spider Activity Reports from Blackwell Synergy") is accessible on the Liblicense listserv at: http://www.library.yale.edu/~llicense/ListArchives/0606/msg00000.html.

3. Barbara Schader, "Case Study in Claiming/Troubleshooting E-Journals: UCLA's Louise M. Darling Biomedical Library," *E-Serials Collection Management: Transitions, Trends, and Technicalities,* ed. David C. Fowler (New York: Haworth Press, 2004), 150.

4. Louise Cole, "A Journey into E-Resource Administration Hell," *Serials Librarian* 49, no. 1/2 (2005): 144-145.

5. Maria Collins, "The Effects of E-Journal Management Tools and Services on Serials Cataloging," *Serials Review* 31, no. 4 (2005): 292-293.

6. Schader, "Case Study in Claiming," 151.

7. Amanda Yesilbas, "Old Is New Again: Using Established Workflows to Handle Electronic Resources," (presentation given at the Twenty-First Annual NASIG Conference, Denver, Colorado, May 4-7, 2006).

8. Ibid.

9. Innovative Interfaces, Inc., "Serials E-Checkin," a product brochure accessible at http://www.iii.com/pdf/lit/eng_echeckin.pdf.

10. Stephen Bosch et al., "Publisher Packages in the E-World: New Roles for Libraries, Publishers, and Agents" (panel discussion given at the 2006 Annual ALA Conference, New Orleans, Louisiana, June 24, 2006).

11. Ibid.

Chapter 17

Issues, Changes, and Trends in Cataloging E-Journals

Bonnie S. Parks

INTRODUCTION

The mid-1990s sparked a huge increase in the popularity of the Internet and consequently, the popularity of e-resources. Initially they were novelties, but it soon became apparent that they would change the way research was conducted, and in turn, force libraries to reevaluate how they provided access to information. Now, more than ten years later, e-resources, specifically e-journals, have become a ubiquitous part of library collections. As a result of the migration from print resources to their electronic counterparts, serials catalogers have been faced with many new and pressing challenges in their mission to provide users with accurate and timely access to information. Among these challenges are the inadequacy of the existing cataloging code, treatment concerns related to multiple versions of a single work, the emergence of new (and hopefully improved) standards to respond to user needs, and the quest for efficiency. This chapter examines the cataloging community's recent efforts to ease the transition from print to e-resources.

CATALOGING E-JOURNALS: ISSUES AND SOLUTIONS

Inadequacy of Cataloging Rules

The cataloging code provides detailed instructions for describing bibliographic resources. However, when presented with the task of

Managing the Transition from Print to Electronic Journals and Resources

applying cataloging rules to electronic serials (hereafter referred to as e-journals), catalogers face more than a few problems that these rules do not effectively address. Indeed, by their nature, serials are dynamic resources, often described as moving targets. When this characteristic is paired with the dynamic nature of the Internet, it becomes clear that serials catalogers face a rather daunting task in their effort to provide users with access to e-journals.

The desire to make e-journals available in library catalogs compelled the cataloging community to reexamine the existing code with respect to the concept of seriality. Earlier versions of the Anglo-American Cataloguing Rules divided publications into two categories: monograph and serial, with serials being defined as, "a publication in any medium issued in successive parts bearing numeric or chronological designations and intended to continue indefinitely."[1] The restrictiveness of the definition did not translate well into the newly emerging Web-based environment, for the dynamic nature of Web sites and databases could not be adequately accommodated by the existing definition. Jean Hirons and Les Hawkins addressed this issue at the North American Serials Interest Group (NASIG) Conference in 1999 where they presented an overview of the proposals contained in *Revising AACR2 to Accommodate Seriality: Report to the Joint Steering Committee for Revision of AACR2*. They noted that

> One of the most characteristic aspects of databases and Web sites is their dynamic nature; they are neither finite nor do they have discrete successive parts. Rather, they are "continuing"; i.e., subject to ongoing updating and/or revision but not necessarily through the addition of discrete parts.[2]

They went on to discuss the concepts of "continuing" and "finite" with respect to bibliographic resources and proffered a revised, more relaxed definition of a serial, "A bibliographic resource issued in a succession of discrete parts, usually bearing numeric and/or chronological designations, that has no predetermined conclusion."[3] The term "usually" was added in relation to the numbering, allowing the inclusion of a serial that lacks designation on the first issue, and the term "discrete" was added to help distinguish from the newly coined "integrating resources." The emergence of e-journals has highlighted other inadequacies of the cataloging code.

For example, two of the key issues in the cataloging of e-journals are (1) how to describe them, and (2) whether to utilize a single-record approach or provide separate records for the electronic and print versions. Very few e-journals are strict reproductions of their print counterparts. While the intellectual content may be identical, they are surrounded by other information which may include publisher information, a journal home page, and an archives page, just to name a few. Amidst these additional features, it can become difficult for catalogers describing e-journals to determine the earliest issue to be used as the basis of description and the appropriate chief source of information.

Rule 12.0B1 of the *Anglo-American Cataloguing Rules,* 2nd Edition (AACR2) instructed catalogers to describe a print serial based on the first or earliest issue in hand. While generally considered straightforward practice when describing a print resource, there are a few problems to take into account when dealing with e-journals. First, e-journals do not always have easily discernable issues. Take, for example, the journals published by BioMed Central, which make articles accessible online immediately following peer review.[4] With these journals, there is no formal concept of an "issue" as a collection of articles packaged for publication; rather each volume is comprised of individually numbered articles with each article constituting an issue. So, while the beginning designation (362 field) for the journal *Biomedical Digital Libraries* looks like this:

362 0# $a 1:1 (20 Sept. 2004)

the second numeral "1" refers not to the first issue, but rather the first article. The same holds true for the date, September 20, 2004, which represents the publication date of the first article.

However, what if the e-journal has a print counterpart, and the first issue is not available electronically? If a cataloger did not have the first or earliest issue available at the time of cataloging, the cataloger was instructed to give a "description based on" note if the online version does not begin with the first issue of the printed version. Publishers, when digitizing long runs of print titles, typically make available more recent issues and work their way backward. Accordingly, it is often the case that the basis of description is not necessarily the first issue even though it is the earliest issue available. To address this problem, the *CONSER Cataloging Manual* (CCM) at one time allowed catalogers to use a "coverage as of" note to imply that, while the cataloger had the earliest issue, there were still earlier issues that remained to be digitized.[5] Unfortunately, this was not a viable long-term solution. Even with a "coverage as of" note, the 362 field required constant maintenance until the first number of the first volume had been digitized. Serials catalogers are used to the ongoing maintenance involved with serials, but continued record maintenance as each back issue is digitized is impractical.

Identifying the Chief Source of Information in an E-Journal

Another common problem with cataloging e-journals had to do with identifying the chief source of information, which is content from which catalogers are instructed to transcribe the title and statement of responsibility. For print serials, AACR2 rule 12.0B2a stated that the chief source of information was the title page or title page substitute.[6] However, e-resources do not have tangible title pages, and in some instances it is difficult to find a formal presentation of the title anywhere on the resource's Web site. Prior to the 2002 revisions, there was no specific rule indicating which issue the description should be based on. As a result, multiple bibliographic records representing the same publication were often created. Web pages for remote access e-resources often contain multiple, granular layers of content. Title presentation in any or all of these layers may vary, thereby making it difficult to select a title proper.

One Record or Two? Addressing Multiple Versions of an E-Journal

The number of bibliographic records used to describe e-journals is another issue for serials catalogers to consider. As was mentioned earlier, many e-journals are online versions of print titles. Some are issued simultaneously, some are republications of earlier print titles, and some abandon the concept of "issue" and publish on an article-by-article basis. E-journals can be issued by the original print publisher, a third party, or even as part of a library's digitization project.

So what does a cataloger do if the library has access to both a print and electronic format of a given journal? Should the library adopt a single-record approach or maintain separate records for the print and the electronic versions of the title?

CONSER policy for cataloging e-journals allows for two approaches to answering these questions. The single record approach provides access to the electronic version of a serial if a library already has the print version in its catalog. The single record approach merely notes the existence of an online version of the print version which can be achieved by adding a limited number of fields to the record for the print title. Following the single record approach, Module 31 of the *CCM* instructs the cataloger to

- Note the availability of the online version in field 530 [Additional Physical Form Available Note]
- Add a 740 (second indicator blank) Title Added Entry or 7XX author/title added entry when the title of the online version differs
- Provide the location [URL] of the online version in field 856
- If a separate International Standard Serial Numbers (ISSN) has been assigned to the online serial but a separate record does not exist, add field 776 [Additional Physical Form Entry]
- Optionally, an e-resource 007 [Format] field may be added for the online version[7]

Maintaining separate records is an option if a library does not subscribe to the print version or if the library prefers to maintain consistency in treatment as with other formats (e.g., film and fiche). Following the separate record approach, *CCM*, Module 31 instructs the cataloger to:

In the record for the original (i.e., print version of the serial):

- note the availability of the online version in field 530 [Additional Physical Form Available Note];
- add a 730 Title Added Entry or 7XX author/title added entry when the title of the online version differs;
- link to the online record with field 776 [Additional Physical Form Entry];
- provide the location [URL] of the online version in field 856 (if not already present in the record)

In the record for the online version of the serial:

- describe the online version using all appropriate fields;
- add a 730 Title Added Entry or 7XX author/title added entry when the title of the original differs;
- link to the original version's record using field 776 [Additional Physical Form Entry];
- give appropriate 856 [URL] fields.[8]

While either of the approaches described in previous text serves as a viable solution, each is not without its drawbacks. Using separate records when purchasing record sets from commercial MARC record providers (Serials Solutions, MARCit!, etc.) makes adding and deleting records more efficient, although this approach may increase the overall number of records for titles in the catalog and confuse users. Using a single-record approach may facilitate the catalog's usability, but record set maintenance may become an issue. Whatever strategy a library chooses to adopt ultimately depends on the size of the collection, the library's policy with respect to other multiple format issues (i.e., microform, CD-ROM), the number of staff, and, of course, local politics.

Aggregator-Neutral Records for E-Journals

E-journal content can be packaged in many ways. One common method is via an aggregator, which, for the purposes of this chapter, describes any "company that provides digitized access to the content of many different serials and other resources, often from a variety of different publishers."[9] Oftentimes, the same serial titles are made available by various aggregators in multiple packages. Creating a separate record for each aggregation can be time-consuming, difficult to maintain, and confusing to users. In July 2003, CONSER libraries agreed to a new set of criteria for cataloging e-journals. Deemed the "aggregator-neutral" approach, the goal of the standard is to simplify the maintenance of e-journal records while "still providing the type of catalog access that serial records have always provided."[10] In addition, it should reduce some of the problems associated with cataloging e-journals by directing catalogers to maintain a single electronic serial record that represents all of the online manifestations of a single electronic serial title. The aggregator-neutral record is a "record that is separate from the print that covers all versions of the same online serial on one record."[11]

The aggregator-neutral record applies to online serials distributed by one or more providers and appears similar to other e-journal records. If the resource contains a uniform title, it will have the same qualifier as the print plus the word "Online." Notes relating to specific packages are not made; however, records may contain multiple URLs representing packages that contain the complete serial. Steve Shadle points out that the aggregator-neutral record also brings into line ISSN Network and CONSER practice, thereby facilitating the use and adoption of records between the two systems.[12] Moveover, Naomi Young notes that utilizing this approach may be viewed as stepping back from the AACR2 idea of cataloging the item in hand, but may be a step toward the kind of cataloging described in the Functional Requirements for Bibliographic Records (FRBR).[13] FRBR (which will be discussed later in this chapter) is the product of a study undertaken following the 1990 Stockholm Seminar on Bibliographic Records "to delineate in clearly defined terms the functions performed by the bibliographic record with respect to various media, various applications, and various user needs."[14] FRBR does so by means of a conceptual model that identifies and defines: (1) entities of interest to users of bibliographic records; (2) their attributes; and (3) the relationships that operate between them.[15]

CONSER Standard Record (formerly Access Level Record)

Yet another cataloging option for e-journals is the CONSER Standard Record, formerly known as the access level record. The development of an access-level MARC/AACR2 catalog record for monographic and integrating remote access e-resources was proposed under the Library of Congress' FY 2003/2004 Strategic Plan.[16] Attendees at the 2005 CONSER/BIBCO Joint Operations meeting supported a wider application of this proposed record, specifically to e-journals. Discussion ensued further at the subsequent CONSER Operations meeting, and the concept for a single standard for serials cataloging was suggested.

The pilot project built upon Tom Delsey and David Reser's work on developing and implementing an access level record for e-resources. Regina Reynolds of the Library of Congress and Diane Boehr of the

National Library of Medicine were appointed cochairs of the project while Ed Jones of National University adapted Delsey's set of user tasks to accommodate serials.[17] The pilot project sought to accomplish three objectives: functionality, cost-effectiveness, and conformity to current standards. The Access Level Record for Serials Working Group was charged with the development of a single CONSER standard record that would apply to both print and online formats, replacing existing multiple record levels (i.e., full, core, and minimal), and reducing serials cataloging costs by requiring in serial records only those elements that are necessary to meet FRBR user tasks (find, identify, select, and obtain).[18]

The group's Final Report notes that cost savings and user benefits also may be realized by recording the elements in a way that is more straightforward for the cataloger to provide and easier for the end user to understand. The report also points out that the emphasis of the access level record is on access points rather than description and that the record is intended to be a "floor" to which additional elements may be added if considered essential to meeting FRBR user tasks for a specific resource or if they meet the needs of a given institution.[19] The pilot project proved successful and implementation of this new record (now called the CONSER Standard Record) will take place no earlier than May 2007.[20]

REVISIONS TO THE CATALOGING RULES

2002 AACR2 Revisions

The 2002 revisions to AACR2 and subsequent revisions of CONSER documentation were two early efforts that sought to address and codify the transitory nature of e-resources. These revisions did indeed provide solutions to several common descriptive problems, including the two problems described in previous text, finding the earliest issue to be used as the basis of description and determining the appropriate chief source.

Under the revised AACR2 guidelines, rule 12.0B1a instructs catalogers to base the description on the first or earliest available issue and to prefer the first or earliest issue over a source associated with the whole serial or with a range of more than one issue or part.[21] This is a change for e-journals. Prior to the revisions, Chapter 12 failed to specify which issue served as the basis of description for nonprint serials. It now clarifies that the preferred source for the title is the first or earliest issue.

But what about those print journals with electronic counterparts that have yet to be fully digitized? Fortunately, the revised guidelines provide a solution to this problem as well. As previously mentioned, when cataloging an online version of a print serial, catalogers are instructed to give a "description based on" note if the online version does not begin with the first issue of the print version. The "description based on" note, however, does not always provide the best solution. Since providers vary in the range of issues that they offer online, the CONSER practice of giving a "coverage as of" note was discontinued when Library of Congress Rule Interpretation (LCRI) 12.7B10 was deleted in 2003.[22] The beginning dates of the print version may be given in a 362 1# field to provide justification for the fixed field beginning date:

362 1# $a Print began with: Vol. 3, no. 1 (Jan. 1984).

Identifying the chief source of information is a crucial first step when cataloging an e-journal. Serials catalogers have long been known to follow the rules in Chapter 12 of AACR2 when describing serials. Catalogers describing e-resources have been instructed to utilize the rules in Chapter 9. With the 2002 amendments to AACR2, catalogers describing e-journals are now directed to use both AACR2 chapters 9 and 12 when identifying the chief source. For Internet serials, catalogers are directed to follow 9.0B1, which states that "the chief source of information of e-resources is the resource itself." The title proper should be taken from formally presented evidence with the source of the title proper being the most complete formal presentation of the title associated with the first or earliest issue.[23] Unlike print serials, electronic versions of serials do not always have a clearly defined title page or cover—rather they have many places where title information may be provided. Common sources of title information include journal home pages, tables of contents pages, title bar displays, "about" pages, digitized versions of journal covers, and even archives pages. So, according to the revised rules, the source of title proper for a remote-access e-journal should be the most complete presentation of title information (AACR2 9. 0B1) in conjunction with the first or earliest available issue (AACR2 12.0B1). Furthermore, AACR2 9.7B3 instructs catalogers to always give the source of the title proper; for example, the source of the title proper for the journal *Biomedical Digital Libraries* would look like this:

500 ## $aTitle from issue caption
(Biomedical Digital Libraries site, viewed Jan. 3, 2006).

Reflecting the fact that cataloging is a cooperative effort, this revised rule regarding the citation of the source of the title proper aims to facilitate future record maintenance efforts.

RESOURCE DESCRIPTION AND ACCESS (RDA)

Certainly the 2002 revision of AACR2 made some headway in addressing description issues related to the transient nature of e-resources. However, applying rules to e-resources that were designed for the description of print resources is like using duct tape to patch a dam; sure, duct tape might be considered a universal fix all, but it will not keep the dam from eventually bursting. Indeed, while there are several tools aside from AACR2 and its corresponding LCRIs, especially LCRI 1.0, to assist catalogers with treatment decisions for e-resources, the fact remains that the development and implementation of the Anglo-American Cataloguing Rules predates the existence of remote-access e-resources, and catalogers currently find themselves in an environment dominated by e-resources. The cataloging community has been aware of this problem for more than a decade. The late-1990s were filled with discussions regarding AACR2's inadequacy in describing Internet resources. For example, Arlene Taylor's article, "Where Does AACR2 Fall Short for Internet Resources?" published in 1999, argued that it may be time to "move the principles of AACR2 beyond their tie to traditional publishing methods."[24] Moreover, Anne Copeland noted

> With electronic manifestations taking on new and unusual characteristics, adhering to cardinal AACR2 rules became problematic. Two discussions were taking place in the literature and overlapped organically: how to work with the current codes and

standards to describe ejournals given their inadequacies, and how to revise the code to address the reality of continuing and integrating electronic resources.[25]

In response to the inadequacy of AACR2 to address the challenges of e-resources, a new standard is under development that will replace AACR2. Originally called AACR3, but then changed to *Resource Description and Access* (RDA), work on this standard began in 2004 and is headed by the Joint Steering Committee for Revision of AACR (JSC), which is working in cooperation with the RDA editor. RDA is described as "a new standard for resource description and access designed for the digital world."[26] It is worth noting that the term "digital world," in this case refers not only to the JSC's vision for including guidelines and instructions for description and access for all digital and analog resources, but also that records created using RDA will be used in a variety of digital environments (the Internet, Web Online Public Access Catalogs [OPACs], etc.).[27] In addition, this term reflects the fact that *RDA* itself will be primarily a digital product (although a print version will also be available). Ideally, RDA will provide a flexible framework for describing both analog and digital resources so that data describing these resources is readily adaptable to new and emerging database structures while remaining compatible with existing records in online library catalogs. The JSC intends to have RDA ready for release in early 2009.[28]

THE INFLUENCE OF FRBR
ON THE NEW CATALOGING CODE

In 1998, the International Federation of Library Associations and Institutions' (IFLA) Study Group on Functional Requirements for Bibliographic Records (FRBR) developed a user-centered, entity-relationship model as a generalized view of the bibliographic universe. This model, known as FRBR, was designed to be independent of any existing cataloging code.[29] FRBR is a theoretical bibliographic model that identifies four specific user tasks ("find," "identify," "select," and "obtain") and recommends basic requirements for bibliographic records. In addition, FRBR conceptualizes three groups of entities. Group One entities consist of the products of intellectual or artistic ventures (e.g., publications). Group Two entities are those corporate bodies or persons responsible for the intellectual content and maintain a specific relationship with Group One entities. Group Three entities are the subjects of works and can be concepts, objects, events, or places.[30]

The internal subdivision of Group One entities is of particular importance. FRBR specifies that intellectual or artistic products include the following types of entities:

- the work, a distinct intellectual or artistic creation;
- the expression, the intellectual or artistic realization of a work;
- the manifestation, the physical embodiment of an expression of a work; and
- the item, a single exemplar of a manifestation.

FRBR also specifies particular relationships between classes of Group One entities. It is important to note that the expression, manifestation, and item entities exhibit a recursive relationship; that is, each is dependent upon the entity going before:

- a work is realized through one or more expressions;
- each of which is embodied in one or more manifestations;
- each of which *is exemplified by* one or more items.

These concepts are significant because FRBR is the basis for Part A (Description) of RDA. RDA will include FRBR terminology when appropriate (e.g., use of the names of the Group One entities: "work," "expression," "manifestation," and "item") and will incorporate the FRBR objectives of allowing users to find, identify, select, and obtain resources.[31]

While the overall application of FRBR principles to RDA is viewed as a step forward in creating a more relevant and applicable cataloging code, it is certainly not without its challenges. It has been argued that continuing resources, more specifically serials, represent the greatest challenges to the application of the FRBR concept. Frieda Rosenberg and Diane Hillman's report "An Approach to Serials with FRBR in Mind : CONSER Task Force on Universal Holdings" (Rev. 1/24/04)[32] addresses several weaknesses of the FRBR model when applied to serials, as does Selden Lamoureux's report "FRBR and Serials: A Complicated Combination."[33] The many questions raised when applying the FRBR model to e-journals include whether different formats of a serial should be given the same record, which entity represents the work (the entity to be cataloged), whether a segment of a run of issues should be identified by one title or name-title, or the entire run of issues associated through time, and to which level of FRBR entity should a serial's ISSN be assigned.

Kristin Antelman argues that study of the bibliographic work is yet to conceptualize and define the serial work, and that it is a misfit in the FRBR model. She further points out that we need to place a greater emphasis on relationships among abstract entities and less on identification of the physical item.[34] Others argue that the FRBR group one entities are readily applicable to serials. Tom Delsey notes that in the FRBR model, the work, expression, and manifestation entities are all clearly applicable to continuing resources.[35] While it seems easy to get caught up in and confused by the details of FRBR, it is important to realize that the utilization of this theoretical model is still in a state of experimentation with potential applications yet to be discovered or fully explored.

CONCLUSION

The challenges discussed in this chapter are just a few of many that the cataloging community faces as libraries transition from print to e-resources. Reexamining the existing cataloging code, addressing the unique descriptive needs of e-journals, and the treatment concerns related to multiple versions of a single work will keep our proverbial plates full for the foreseeable future. New discovery and retrieval tools are forcing librarians to reevaluate the way in which catalogs provide information to users. In the quest for new (and hopefully improved) standards to respond to user needs, catalogers must ensure that library catalogs describe and provide access to not only a particular item, but also intellectual works and the manifestations of those works. This is certainly a time of transition, and, while it remains to be seen whether the cataloging community can successfully integrate the aforementioned changes into cataloging practices, it is important that flexibility and a sense of humor be maintained. As information professionals redefine their roles, their practices and even their standards to meet the challenges of e-resource cataloging, they must also keep the information needs of users at the forefront of their minds.

NOTES

1. Anglo-American Cataloguing Rules (AACR), 2nd ed., 1998 rev.

2. French, Pat. "AACR2 and You: Revising AACR2 to Accommodate Seriality." *The Serials Librarian,* 38 (3/4) 2000, 251.

3. Ibid.

4. BioMed Central information page, http://www.biomedcentral.com/info/, accessed December 9, 2006.

5. CONSER Cataloging Manual, Module 31.

6. AACR, 2nd ed., 1998 rev.

7. CONSER Cataloging Manual, Module 31.

8. Ibid.

9. CONSER Cataloging Manual, Module 31, p. 44.

10. Steve Shadle, "The Aggregator Neutral Record: Putting Procedures into Practice," *The Serials Librarian* vol. 47(1/2) 2004, 151.

11. FAQ on the Aggregator-Neutral Record, http://www.loc.gov/acq/conser/agg-neut-faq.html, May 29, 2003, accessed June 2, 2006.

12. Shadle, 152.

13. Naomi Young. "The Aggregator-Neutral Record: New Procedures for Cataloging Continuing Resources." *The Serials Librarian* vol. 45 (4) 2004, 42.

14. Jones, Ed. "Summit on Serials in the Digital Environment." Glossary. http://www.loc.gov/acq/conser/glossary.html, accessed March 20, 2007.

15. Ibid.

16. Cataloging Directorate Strategic Plan, Goal 4, Group 2: Processing Rule Analysis Group Report, http://www.loc.gov/catdir/stratplan/goal4wg2report.pdf, accessed October 7, 2007.

17. Access Level Record for Serials [Charge], (http://www.loc.gov/acq/conser/pdf/ChargeIc-pccaug17.pdf, accessed December 1, 2006.

18. From: Access Level Record for Serials Objectives, draft, updated 1/17/06, http://www.loc.gov/acq/conser/pdf/WorkingGroupobjectivefinal.pdf, accessed June 2, 2006).

19. Access Level Record for Serials Working Group: Final Report, http://www.loc.gov/acq/conser/alrFinalReport.html, accessed December 7, 2006.

20. Statement by Mecheal Charbonneau, Program for Cooperative Cataloging Policy Committee Chair made on Jan. 26, 2007, http://www.loc.gov/catdir/cpso/conserdelay.html.

21. Ibid.

22. CONSER Cataloging Manual Module 31, p. 29, http://www.loc.gov/acq/conser/Module31.pdf, accessed November 15, 2006.

23. AACR2 2002 Rev. ed.

24. Arlene G. Taylor, "Where Does AACR2 Fall Short for Internet Resources?" *Journal of Internet Cataloging* 2, no. 2 (1999): 43-50.

25. Anne Copeland, "E-Serials Cataloging in the 1990s: A Review of the Literature," *The Serials Librarian,* 41, no. 3-4 (2002): 16.

26. RDA FAQ, http://www.collectionscanada.ca/jsc/rdafaq.html, accessed December 9, 2006.

27. Ibid.

28. Ibid.

29. Tillett, Barbara, "What is FRBR: A Conceptual Model for the Bibliographic Universe," Library of Congress Cataloging and Distribution Service, 2004, p. 2.

30. Ibid., p. 3.

31. Joint Steering Committee for Revision of AACR: RDA, http://www.collectioncanada.ca/jsc/rda.html, accessed December 1, 2006.

32. Rosenberg, Freida and Hillman, Diane, "An Approach to Serials with FRBR in Mind : CONSER Task Force on Universal Holdings" (Rev. 1/24/04).

33. Selden Lamoreaux in FRBR and Serials: "A Complicated Combination" [on line]. Chapel Hill, NC: The University of North Carolina at Chapel Hill, 2004. http://www.unc.edu/~lamours/papers/FRBR.doc, accessed Dec. 1, 2006.

34. Antelman, Kristin. "Identifying the Serial Work as a Bibliographic Entity." *Library Resources and Technical Services*, 48 (4) 2004.

35. Delsey, Tom, "FRBR and Serials," http://www.ifla.org/VII/s13/wgfrbr/papers/delsey.pdf, accessed March 21, 2006.

Chapter 18

Workflows for Managing E-Resources: Case Studies of the Strategies At Five Academic Libraries

Patrick L. Carr

INTRODUCTION

Earlier chapters have demonstrated that making the transition to e-resource oriented workflows is a significant challenge requiring creativity, collaboration, and the ability to embrace change. Unfortunately, there is not one "correct" set of workflows that all libraries can implement in order to overcome this challenge. Instead, each library must develop its own unique solutions that are shaped by such factors as its size, resources, user community, and organizational structure. By examining the e-resource management workflows of peers, however, a library can gain valuable insights into the solutions that will most effectively meet its own needs. This chapter aims to provide such insights through case studies describing the workflows for e-resource management developed by five academic libraries. The first section of the chapter provides a broad overview of the case libraries, highlighting their unique characteristics and overall e-resource management strategies. The next section builds upon this overview by describing how the libraries perform core tasks for e-resource management. The chapter concludes by discussing a few overall insights to be gained from an analysis of the case libraries' workflows.

AN OVERVIEW OF THE LIBRARIES STUDIED

The five academic libraries described in this chapter were contacted in the fall of 2006. During this time, representatives from each library were questioned about e-resource management workflows through phone conversations and follow-up e-mail exchanges. The participating libraries were selected due to their differing sizes and approaches to meeting the challenges of e-resource management. This section and Table 18.1 give an overview of the unique characteristics of each

TABLE 18.1. Overview of the Case Libraries

Institution and Appr. 2004/2005 Enrollment	Library Home Page	Carnegie Classification	Respondents
Central Piedmont Community College (Charlotte, North Carolina) **56,805**	http://www1.cpcc.edu/library/	Associate's-Public Urban-Serving Multicampus	Jennifer Arnold (Senior Librarian)
The College of New Jersey (Ewing, New Jersey) **6,812**	http://www.tcnj.edu/~library/index.php	Master's Colleges and Universities (larger programs)	Jia Mi (Electronic Resources/Serials Librarian)
Georgia Institute of Technology (Atlanta, Georgia) **16,841**	http://www.library.gatech.edu/	Research Universities (very high research activity)	Bonnie Tijerina (Electronic Resources Coordinator, Collection Development) and Elizabeth L. Winter (Electronic Resources Coordinator, Acquisitions)
Haverford College (Haverford, Pennsylvania) **1,172**	http://www.haverford.edu/library/	Baccalaureate Colleges-Arts & Sciences	Norm Medeiros (Associate Librarian of the College and Coordinator for Bibliographic and Digital Services) and Marilyn Creamer (Serials Specialist)
University of Nevada, Reno (Reno, Nevada) **15,950**	http://www.library.unr.edu/	Research Universities (high research activity)	Paoshan W. Yue (Electronic Resources Access Librarian) and Rick Anderson (Director of Resource Acquisition)

library and the impact of these characteristics on their overall strategies for e-resource management.

Central Piedmont Community College

As an institution with six campuses in North Carolina and sustained growth in its distance education programs, a driving force in the Central Piedmont Community College (CPCC) Library's transition to acquiring resources in electronic formats has been the need to provide users with access to these resources regardless of their geographic location. Although accreditation requirements have resulted in the library maintaining print subscriptions to approximately three to four hundred of its journals, the print collection has shrunk dramatically in favor of online only subscriptions and will continue to shrink in the future. In order to develop and carry out the workflows that will make this transition a success, CPCC's senior librarian responsible for serial and Electronic Resource Management (ERM) has worked closely with librarians who have responsibilities focused on systems and Web services. An additional partner for the library is CPCC's Information Technology Services unit, which has been crucial in ensuring that the institution's EZproxy server is configured to provide all appropriate portions of the library's widely dispersed user community with online access to its e-resources.

The College of New Jersey

Workflows for e-resource management at The College of New Jersey (TCNJ) Library center around one person—Jia Mi, the electronic resources/serials librarian. Created in 2003, the electronic resources/serials librarian position was developed due to the library's decision that, as its e-resource collection continued to grow, the successful management of this collection could no longer be divided between electronic databases and electronic serials. As a result, the library created a new position within its Public Services Department that oversees e-resource and serial collection development, license negotiation, budget management, activation, and maintenance. This position centralizes all e-resource management responsibilities under the umbrella of one individual who has an opportunity to develop an understanding of the overall workflows and who can collaborate closely with other information professionals in the library in order to ensure that each component of these workflows runs effectively. Moreover, as Mi indicates in a 2005 National Serials Interest Group Conference (NASIG) conference presentation, the fact that the electronic resources/serials librarian is a public services position and has reference responsibilities gives this professional the ability to develop a broad perspective on how e-resource management workflows function to meet user needs.[1]

Georgia Institute of Technology

As Tyler Walters discusses in a 2006 Electronic Resources & Libraries conference presentation,[2] the Library and Information Center at the Georgia Institute of Technology (GT) began in-depth discussions of e-resource management strategies in 2002, the year in which this Association of Research Libraries (ARL) library completed a five-year strategic plan for making the transition from collecting print resources to providing access to e-resources. Also in 2002, the GT Faculty Senate passed a resolution supporting the library's efforts to shift its resources to electronic platforms. Shortly after these events, the library changed the name of its technical services division from "Technical Services & Systems" to "Technology Resources & Services" and outlined a new mission statement for the division that positions e-resource acquisition, organization, access, and preservation at the center of its workflows. To succeed in this transition, the library relied upon the synergy generated from small teams of three to six information professionals. Moreover, in order to facilitate an environment of teamwork, collaboration, lateral communication, and independence in decision making, the library flattened its hierarchy of reporting among the various units in its Technology Resources & Services division. It also created two new positions to spearhead its management of e-resources. The first of these positions, e-resource coordinator for collection development, is responsible for such tasks as selection, budgeting, usage, promotion, evaluation, and close collaboration with subject librarians. The other position, e-resource coordinator for acquisitions, is responsible for such tasks as licensing, invoicing, activation, access troubleshooting, and preservation. Together, the two coordinators play a key role in fostering a cross-disciplinary and cross-functional team of information professionals that organically developed among personnel from such departments as

Collection Development, Acquisitions, Systems, Cataloging, and Information Services.

Haverford College

Haverford College (HC) Library's transition from maintaining a collection of print resources to providing online access to e-resources began in the late 1990s when the library's user community—particularly the faculty in the college's science departments—began to strongly urge the library to provide online access to its resources. Since this time, a number of factors—including the implementation of the SFX OpenURL link resolver and the price savings of online only subscriptions—have prompted the library to rely heavily upon serving its users' information needs through online access to e-resources. A crucial component in this transition has been HC's partnership with the libraries at two other small, liberal arts colleges in southeastern Pennsylvania: Bryn Mawr College and Swarthmore College. Together, these three libraries form the Tri-College Library Consortium, a partnership which provides a framework through which the libraries collaborate to carry out many of their e-resource management workflows including product evaluation, license negotiation, resource activation, and—through Tripod, the libraries' shared catalog—access maintenance. HC's coordinator for bibliographic and digital services plays a leading role in administering the library's e-resource collection development and workflows. Once the consortium has acquired an e-resource, the library's Serials Specialist carries out many of the e-resource management responsibilities.

University of Nevada, Reno

As Yue and Anderson discuss in an article published in *The Serials Librarian*,[3] the year 1998 marked the point in which the University of Nevada, Reno (UNR) Libraries began acquiring e-journals. Since this time, the library has established a policy of subscribing to resources in electronic form whenever such a subscription is possible and licensing terms are acceptable. As this policy has transformed the UNR collection, the library has strived to transform positions and workflows in order to effectively manage e-resources. In 2002, this effort has resulted in the development of a new professional position, electronic resources access librarian, and the establishment of a small unit within the library's Catalog Department in which this position as well as two staff positions work toward ensuring that access is maintained to subscribed e-resources. In addition, important components in the library's e-resource management workflows are carried out by its director of resource acquisition and personnel within its Serials Department, Systems Department, Acquisitions Department, and branch libraries. Prompted by rapid changes in personnel and the growing ambiguities of e-resource management, the library embarked in 2005 on developing a detailed flowchart depicting tasks and responsibilities related to all aspects of e-resource management.[4]

CORE TASKS FOR E-RESOURCE MANAGEMENT

Although workflows for managing e-resources are diverse, most of the core tasks that these workflows aim to achieve are common to all libraries. Appendix B of *Electronic Resource Management,* the highly influential report issued by the Electronic Resources Management Initiative (ERMI) of the Digital Library Federation (DLF) in 2004, includes a detailed flowchart illustrating the complex web of tasks and communications that workflows for e-resource management entail. Figure 18.1, which reproduces the report's overview of this flowchart, identifies the basic tasks that a library must carry out following its notification of a new e-resource. This section describes how the case libraries complete these tasks.

Product Consideration

Product consideration includes all actions that a library must perform in order to determine whether it will pursue a subscription (or some other means of access) to an e-resource. These actions may include evaluating the resource in light of user needs, investigating or developing a consortial partnership for access, and administering a product trial. Three of the case libraries (CPCC, HC, and UNR) rely upon specific committees to spearhead their evaluation of an e-resource. At CPCC, product evaluation is a responsibility of its Collection Development Committee, which consists of information professionals at the library who meet on a monthly basis in order to address collection development-related challenges and assess new resources. At UNR

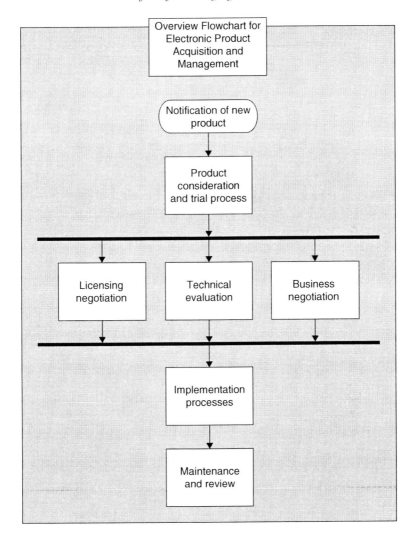

FIGURE 18.1. Overview Flowchart for E-Resource Management (*Source:* Appendix B of the 2004 DLF ERMI report, Electronic Resource Management.)

there is a similar arrangement; here, an e-resource's evaluation is carried out by the library's Collections Working Group, which evaluates all requests for new subscriptions or databases costing more than one hundred dollars per year. Prior to this evaluation, UNR's workflows specify that the library's director of resource acquisition will investigate if a consortial partnership is possible for the acquisition of this resource. As is the case at CPCC, UNR will often administer a product trial for e-resources of any significant subscription cost. The committee responsible for evaluating resources at HC is unique among the case libraries in that it is not a component of the library itself but is instead a component of the partnership mentioned in previous text, the Tri-College Library Consortium. This consortium's Electronic Resources Group, which includes representatives from each of the three partner libraries, evaluates e-resources and makes a collective decision on behalf of the libraries. One member of this committee acts as a trials coordinator who administers trial access for the partner libraries.

The two other case libraries, GT and TCNJ, do not have specific committees for evaluating new e-resources. At GT, such an evaluation is coordinated by the e-resource coordinator for collection development. This position receives all recommendations for new e-resources and communicates with publishers to set up trials, training sessions, and on-site demonstrations. This position also maintains trial blogs, which function as forums through which information professionals at GT can easily share their thoughts and evaluations concerning the e-resources being considered. At TCNJ there is a similar situation in which one position, the electronic resources/serials librarian, receives requests for new e-resources and coordinates the evaluation of these resources. This position's actions include setting up a trial and gathering information about users' needs for the resource. As is the case at GT, although there is no designated committee, information professionals at TCNJ work collaboratively to make collective decisions about whether to pursue access to e-resources.

Negotiations and Technical Evaluation

After a library determines that it wishes to pursue access to an e-resource, Appendix B of the DLF ERMI report cited in previous text indicates that the library must complete three parallel tasks: license negotiations, business negotiations, and technical evaluation. All five of the case libraries integrate license and business negotiations so that they can be simultaneously carried out by a single librarian. Four of the five case libraries (CPCC, TCNJ, GT, and UNR) assign negotiation responsibilities to an internally employed librarian. The primary positions at the libraries with these responsibilities are as follows:

- CPCC: senior librarian responsible for serial and ERM;
- TCNJ: electronic resources/serials librarian;
- GT: e-resource coordinator for acquisitions;
- UNR: director of resource acquisition.

As was the case with product consideration, HC's workflows were unique among the case libraries. One reason for this is that the individual negotiating on the library's behalf is a member of the Tri-College Library Consortium's Electronic Resources Group rather than a librarian employed by HC. In addition, the consortium has developed a model license that it generally presents to the publishers with which it has entered into negotiations. The purpose of the model license is to streamline the negotiation process by having a document in place that already reflects licensing conditions that are acceptable to the consortium.

With respect to the technical evaluation of an e-resource, the case libraries indicate that they have not incorporated a formal component into their workflows through which systems personnel routinely assess the technical feasibility of implementing an e-resource in a process that occurs concurrently with negotiations. Although the case libraries all emphasize that they maintain close partnerships with their respective systems departments, they state that, in the vast majority of their e-resource acquisitions, a technical evaluation is unnecessary. Indeed, most e-resources that the case libraries acquire are accessed via an authenticated Web connection, a routine arrangement in which technical implementation barriers are rare. However, the case libraries do have procedures in place for addressing instances in which there is some degree of uncertainty about the feasibility of a technical implementation. For example, UNR notes that its Systems Department generally conducts a technical evaluation in instances in which the implementation of an e-resource will require the library to install on-site software. In order to save the time of the library's Systems Department, UNR only consults its Systems Department after it has carried out negotiations and is sure that it does indeed wish to acquire the e-resource.

Implementation Processes

A library's acquisition of an e-resource is followed by its implementation, which can include a wide variety of actions such as confirming and recording details of the acquisition, activating the e-resource, customizing its interface, notifying relevant personnel within the library of the access acquired, creating promotional and instructional materials, and preparing for the collection of usage statistics. Thanks to the significant level of detail in its flowchart for e-resource management, UNR provides an effective starting point for discussing the case libraries' implementation processes. Once an e-resource at UNR is ordered and the license agreement is signed, the library's Serials Department notifies cataloging personnel and service administrators of the resource's impending implementation, confirms the completeness of the library's administrative information, and then activates the resource. If installation of the resource is required, the UNR's Serials Department works with the library's Systems Department to ensure that the resource is installed in accordance with its licensing terms. Following these steps, the library's public services librarians work to customize the resource's interface. The workflows described by the other case libraries were similar in nature, requiring librarians with e-resource management responsibilities to closely collaborate with other personnel in the library to activate the resource, make it accessible in the library's e-resource access points, customize its interface, and communicate its accessibility to relevant partners. To a large extent, these processes were spearheaded by the following personnel:

- CPCC: senior librarian responsible for serial and ERM;
- TCNJ: electronic resources/serials librarian;
- GT: e-resource coordinator for acquisitions;
- HC: members of the Tri-College Library Consortium's Technical Services Departments.

With respect to the promotion of an e-resource, the case libraries indicated that they typically rely upon public services professionals and subject specialists to notify their user communities of a new e-resource's accessibility. Venues for promotion that were cited included instruction sessions, posters, and press releases, which could be published on institutional Web sites and in newsletters. Not surprisingly given that it supports the information needs of a user community of approximately 57,000 stretched across a large geographical area, CPCC has a particularly comprehensive promotion program in which it uses newsletter articles, training sessions, and announcements sent to all students' e-mail accounts in order to market the accessibility of new e-resources.

A final component of e-resource implementation is preparing for and collecting usage statistics. Several of the case libraries indicated that in the future they hope to automate the collection of usage statistics through tools and services that leverage the (National Information Standards Organization) NISO Standardized Usage Statistics Harvesting Initiative (SUSHI). Until such tools and services are implemented, however, most of the case libraries have developed workflows in which the collection of usage statistics is supervised by a librarian and carried out by staff members or student workers. At GT, for example, the library's e-resource coordinator for collection development supervises the collection of usage data by a library staff member. Likewise, at HC, the library's coordinator for bibliographic and digital services supervises student workers' ongoing collection and entry of usage data into a master spreadsheet. CPCC and UNR both have similar arrangements for collecting usage data, with the latter library having a procedure in place to issue periodic inquiries concerning the availability of usage statistics from e-resource vendors that currently do not provide this information to customers. Reflecting the relatively small staff at TCNJ, the collection of usage statistics at this library is currently carried out by the library's electronic resource's librarian; formerly, this responsibility was assigned to a graduate student, but the library does not currently have the funds to employ such a student.

Maintenance

The final component of workflows for e-resource management is product maintenance. For the purposes of this chapter, product maintenance includes all of the actions that a library must take to ensure that it maintains access to and accurate records concerning an e-resource that it has acquired and implemented. Reflecting their limited resources and many responsibilities, none of the case libraries indicated that they have developed workflows through which they can systematically verify their continuing access to e-resources (e.g., e-journal check-in). Instead, they indicated a strategy of "putting out fires" through which they work to address and resolve e-resource access problems as these problems are reported by users and personnel.

The case libraries utilize a variety of procedures and tools in order to maintain records concerning licensing terms, access problems, and workflow tracking. UNR relies upon the Electronic Resource Management (ERM) system developed by Innovative Interfaces in order to track administrative metadata concerning e-resources. The population of data into this tool is supervised by the director of resource acquisition and carried out by Serials Department staff members. Another tool central to UNR's management of e-resources is a locally developed Access database, which the library utilizes in order to generate the Web interface that users see when searching the library's A-Z lists of e-resources; this database is populated and maintained by the electronic resources access librarian and her staff.

The four other case libraries (CPCC, TCNJ, GT, and HC) are currently without an ERM system and are relying upon paper files, spreadsheets, the acquisitions module of their Integrated Library System (ILS), and other tools until an ERM system is implemented. Two of these libraries, HC and GT, had previously implemented locally developed tools for tracking their e-resource's administrative metadata. Indeed, HC was actually one of the earliest libraries to implement a tool for ERM. Sponsored by a Mellon Foundation Grant and developed by the Tri-College Library Consortium, the Electronic Resources Tracking System (ERTS) was a Filemaker Pro database implemented in 2002 in order to enable the consortium to manage its administrative metadata. At present, however, the consortium is no longer using this tool. Initially, the consortium had intended to implement a new ERM by participating in a beta project for the development of VERIFY, the ERM system that was developed by Visionary Technology in Library Solutions. Unfortunately, this development project was discontinued and the consortium is currently evaluating ERM systems offered by other commercial vendors. In a scenario similar to HC, GT had previously used a locally developed Access database in order to track administrative metadata concerning e-resources but stopped updating this database in early 2005. GT is currently investigating commercial ERM solutions.

CONCLUSION

Although the specific workflows carried out by the case libraries constitute only a small sampling of the many strategies through which a library can manage its e-resources, the preceding discussion does provide some valuable insights into the challenges libraries face as they develop workflows that reflect the evolving nature of their collections. Among the most significant insights to be gained from studying the case libraries' workflows is the degree to which a library's size, resources, user community, and organizational structure can combine to impact its workflows for managing e-resources. Norm Medeiros, a respondent at HC, points out that the greatest differences between the efforts of small and large libraries to transition to workflows for e-resources are rooted in the fact that small libraries have fewer and less compartmentalized staff members. According to Medeiros, the result is that personnel at small libraries are accustomed to adopting and carrying out a wide variety of responsibilities, a fact that can make the transition to e-resource workflows seem less traumatic.

As is apparent from the changes discussed in previous text, in personnel and organizational structure at GT, a larger library's transition to e-resource workflows can require an extensive reconceptualization of staff responsibilities and departmental goals. With this reconceptualization comes an increased need for communication and collaboration among the staff facilitating e-resource workflows, which is a challenge cited by a number of the case libraries. At GT, for example, one way this challenge is being addressed is through the strong emphasis that the library places on collaboration among personnel and cross-training in order to foster new skills to carry out evolving workflows; accompanying this effort is the library's decision to flatten its hierarchy in order to decrease compartmentalization and increase the flexibility necessary for effective collaboration. In addition to achieving this increased degree of collaboration through face-to-face interaction, GT is striving to integrate the use of collaborative technologies (e.g., Wikis, blogs, and Google spreadsheets) into its workflows for e-resource management. At another of the case libraries, UNR, one way in which communication is being enhanced is through the development of the flowchart cited in previous text detailing e-resource management responsibilities. This flowchart enables staff members involved in e-resource workflows to understand the relationship between their specific responsibilities and the bigger picture of e-resource management at the library.

Looking to the future, the case libraries all expressed an awareness that meeting the e-resource management challenges of their changing environments will require the ability to continue to evolve. As was noted in previous text, one way in which the libraries aim to do this is through the implementation of new tools, such as an ERM system. In implementing these tools, one of the most significant difficulties the case libraries face is identifying tools that meet their unique needs. HC, for example, notes the lack of commercial ERM systems that can effectively meet the needs of a library with e-resource workflows that are rooted in consortial partnership. Likewise, CPCC remarks upon the difficulty of identifying e-resources and e-resource management tools that are appropriate for a community college setting. Along with the implementation of new resources and tools, CPCC indicates that at some point in the future it may need to develop a librarian position which centralizes the library's e-resource management tasks and spearheads relevant projects. In developing such a position, CPCC would follow in the path of other case libraries such as TCNJ, GT, and UNR.

Of course, one of the great challenges in implementing new tools and developing new positions is acquiring the sustained funding needed to support these endeavors. TCNJ, for example, notes that, with users' increasing expectations for access to costly new e-resources, finding funds to allocate to the implementation of tools to manage the library's existing e-resource collection will not be easy. This scarcity of funding reflects the larger challenge that all libraries face in their development of workflows for e-resources: achieving great ends with very limited means. Indeed, as the case libraries' workflows show, making the transition to workflows oriented toward e-resources requires the assessment of a library's limited resources in order to develop strategies that take full advantage of the library's unique strengths.

NOTES

1. Pat Loghry, recorder for a presentation by Jia Mi and Paula Sullenger, "Examining Workflows and Redefining Roles: Auburn University and The College of New Jersey," *Serials Librarian* 50, no. 3/4 (2006): 280.

2. Tyler Walters, "Who Moved My E-Journal: Electronic Resources and Organizational Change," (panel discussion at the 2006 Electronic Resources and Libraries Conference, Atlanta, Georgia, March 2006).

3. Paoshan W. Yue and Rick Anderson, "Capturing Electronic Journals Management in a Flowchart," *Serials Librarian* 51, no. 3/4 (2006) (in press).

4. University of Nevada, Reno Libraries, "UNR E-Serials Work Flow," http://www2.library.unr.edu/serials/ERMworkflow.pdf (accessed December 2, 2006).

5. Timothy Jewell et al., *Electronic Resource Management: Report of the DLF ERMI Initiative* (Washington, DC: Digital Library Federation, 2004), http://www.diglib.org/pubs/dlf102/ (accessed December 2, 2006).

6. Norm Medeiros et al., "Managing Administrative Metadata: The Tri-College Consortium's Electronic Resources Tracking System (ERTS)," *Library Resources & Technical Services* 47, no. 1 (2003): 28.

7. Visionary Technology in Library Solutions, "Tri-College Library Consortium Chooses VERIFY for ERM Solution" (Press Release) http://www.vtls.com/Corporate/Releases/2005/15.shtml (accessed December 2, 2006).

Index